BIOMARKERS

The 10 Determinants of Aging You Can Control

William Evans, Ph.D., and Irwin H. Rosenberg, M.D.,
with Jacqueline Thompson

SIMON & SCHUSTER

New York   *London*   *Toronto*   *Sydney*   *Tokyo*   *Singapore*

Simon & Schuster
Simon & Schuster Building
Rockefeller Center
1230 Avenue of the Americas
New York, New York 10020

DESIGNED BY BARBARA MARKS
Manufactured in the United States of America

5   7   9   10   8   6   4

Library of Congress Cataloging-in-Publication Data
Evans, William, date
Biomarkers: the 10 determinants of aging you can
control/William Evans and Irwin H. Rosenberg, with
Jacqueline Thompson.
p.   cm.
Includes index.
1. Longevity.   2. Aging.   3. Health.   I. Rosenberg,
Irwin H.   II. Thompson, Jacqueline.   III. Title.
RA776.75.E93   1991
613'.0438—dc20   91-9133
CIP
ISBN 0-671-68547-3

ACKNOWLEDGMENTS

Books such as this one build upon the efforts of many people. We wish to acknowledge particularly the contributions of our research colleagues, Drs. Joseph Cannon, Walter Frontera, Miriam Nelson, Maria Fiatarone, and Carol Meredith, and of Virginia Hughes, Elizabeth Fisher, Roger Fielding, Wayne Campbell, and Susan Coehlo. Their dedicated work has contributed much to the scientific basis of this book.

Dietitians Carol Stollar, Rebecca Roubenoff, and Sharon Bortz of Tufts University have helped us to translate the nutritional goals in this book into concrete suggestions.

Yemi, our New York City–based illustrator, has worked hard with us on the exacting details of the drawings.

We are grateful to Sol Gittleman, Provost of Tufts University, Dr. Daphne Roe of Cornell University, Dr. Evan Hadley of the National Institutes of Health, and Dr. Barbara Bowman of Georgia State University for their careful review of the manuscript and for their many helpful suggestions. If errors or misunderstandings remain, they are our responsibility.

Fred Hills at Simon & Schuster has been both creative and patient in his guidance. We are particularly grateful to Herbert Katz, our literary agent, for initiating this project on the basis of his belief that these scientific accomplishments needed to be translated into a form accessible to the public, and for helping us at each stage of the enterprise.

*We dedicate this book to all of those volunteers
who participated in the many studies described in
this book. They have given generously of their
time and energy, in the hope that this work would
contribute to knowledge about exercise, nutrition,
and aging, and thus to the health and welfare of
their fellow men and women. We hope that we
have been worthy of their faith and trust.*

No book can replace the services of a trained physician or exercise physiologist. Any application of the recommendations set forth in the following pages is at the reader's discretion and sole risk. If you are under a physician's care for any condition, he or she can advise you about information described in this book.

CONTENTS

PART IV
STRETCHING YOUR HEALTH SPAN
VIA DIETARY CHANGE

RESOURCES

INTRODUCTION

———

George Burns, America's favorite nonagenarian, complains that too many older people practice playing old.[1] They "think" themselves into their dotage by adopting what they consider to be the expected mannerisms and lifestyle of the elderly. That lifestyle is long on inactivity and semi-dependence and short on vigorous exercise and self-reliance.

Sadly, George Burns is right. Outmoded ideas about aging and what it signifies still pervade our society. Aren't nursing homes often called "rest homes"? The names implies that older people, like Rip Van Winkle, are only capable of one thing—a long, leisurely snooze. This book is dedicated to the goal of erasing such old-fashioned notions from the public consciousness.

Many of the misconceptions about aging are based on studies using animal subjects or on extrapolations from research done with young adults. We at the U.S. Department of Agriculture's Human Nutrition Center on Aging (HNRCA) at Tufts University feel that, to learn more about the biological destiny of aging human beings, *it's critical to study aging human beings.* It makes little sense to try to adapt conclusions based upon animal studies, or even studies of young adults. If the subject is aging, we need to study mature individuals.

This is a book of hope because much of what we've found in our research laboratories in Boston and through clinical studies using mature subjects belies the traditional view of "senescence," which is the scientific term for old age.

For example, we've found that no group in our population can benefit more from exercise than senior citizens, as startling a statement as that may seem at first blush. Indeed, *the muscles of elderly people are just as responsive to weight lifting as those of younger people.*

A 12-week program of strength training using 60- and 70-year-old men resulted not only in substantial increases in strength (their lifting ability went from 44 to 85 pounds), but also in muscles that were larger and leaner with less fat in and around them. An 8-week study of 87- to 96-year-old women confined to a nursing home showed that resistance exercise tripled their muscles' strength and increased their size by 10 percent.

This leads to an important conclusion: *Much of the loss of muscle as we age is preventable—and even reversible.*

The results we just cited are the norm, not aberrations. We've seen the so-called young old increase their strength by almost 200 percent and their muscle mass by some 15 percent; and the old-old group—the frail elderly—increase strength by as much as 180 percent and muscle mass by up to 12 percent. In terms of overall physical function, our elderly study participants regularly experience increases as large as 50 percent.

APPLYING OUR RESEARCH FINDINGS TO THE MIDDLE-AGED

So what does our research on 70-, 80-, and 90-year-old subjects mean to you, our readers, who are likely to be 30 years younger?

What we're saying is best summed up by these assertions in the following two paragraphs:

Advanced age is not a static, irreversible biological condition of unwavering decrepitude. Rather, it's *a dynamic state that, in most people, can be changed for the better no matter how many years they've lived or neglected their body in the past.*

Yes, you do have a second chance to right the wrongs you've committed against your body. Your body can be rejuvenated. You can regain vigor, vitality, muscular strength, and aerobic endurance you thought were gone forever. Based on our research, this is possible whether you're middle-aged or pushing 80. *The "markers" of biological aging can be more than altered: in the case of specific physiological functions, they can actually be reversed.*

REST IS PRECISELY WHAT AGING PEOPLE DO *NOT* NEED

To be sure, when you arrive at age 55 or 60, it is not time to put your feet up and take it easy for the rest of your life. In truth, at no time during your lifetime is putting your feet up and resting for extended periods of time a good idea.

Why?

The answer lies not only in our research at Tufts, but in the parallels other researchers have drawn between extended bed rest and the inactivity that characterizes the lifestyle of many older people. Within the medical community, it is now widely recognized that prolonged bed rest, even for the sick and frail, only makes matters worse. It may surprise you to learn that all the changes that typically occur with age can occur *within a matter of days* in anyone, including young people, who are forced to stay in bed.

In the late 1960s, the Swedish physiologist Bengt Saltin asked five young men to remain in bed 24 hours a day for three weeks to study their bodies' physiological response. Two of the men were athletes and the remaining three ranged from relatively active to sedentary. Saltin found that within those 21 days all of the men experienced a dramatic drop in their aerobic capacity *equivalent to almost 20 years of aging!*[2]

More than a decade later in his well-known essay, "Disuse and Aging,"[3] Dr. Walter M. Bortz II, who also studied the deleterious effects of bed rest, came to the conclusion that *at least a portion of the changes that are commonly attributed to aging are, in reality, caused by immobility. As such, they're subject to correction by mobility—meaning activity and exercise.*

"Through the long eons in which our forebears were physically active as a necessity of survival, they died of starvation, injury, and infection," Bortz observed. "In our current golf-cart age in which two of these major historic killers are largely controlled, we die of degenerative diseases, on which the impact of our physical inactivity may be considerable."

Bortz first made the connection between bed rest and disuse after an Achilles tendon tear when his right leg languished in a cast for six weeks. Bortz was 35 at the time, but he says that when the cast was removed, he was shocked to discover a limb that "had all the appearances of one belonging to a person 40 years older. It was withered, stiff, and painful." He immediately perceived a link between immobility and accelerated aging worth serious investigation.

A thorough search through the medical literature bore out his hypothesis. _A decline in a number of physiological functions characterizes the "decay curve" following both bed rest and insufficient exercise at any age._ Both cause a change in the fat/lean–body mass ratio in favor of fat; a drop in maximum oxygen consumption (VO_2 max.) and cardiac output; a diminished sense of balance; a loss of body water and a lessened ability to control the body's internal temperature; fewer red blood cells and an increased tendency to thrombotic disease (blood clots); a rise in total blood lipids (cholesterol and triglycerides); calcium loss from the bones; and a substantial insensitivity to glucose. In addition, both bed rest and inactivity alter the body's ability to assimilate drugs and have an adverse effect on a person's sense of taste and hearing.

Bortz writes: " 'Use it or lose it' is a pervasive biologic law, the application of which has received insufficient attention where the human body is concerned. Medicine particularly has been slow to recognize the benefits of exercise in a number of disease states."

He concludes, "The point of this survey is not to presume that physical inactivity is the cause of the aging process. It is wrong to suggest that exercise might halt the fall of the grains of sand in the hourglass. It is proposed, however, that the dimension of the aperture may be responsive to the toning influence of physical activity, and consequently the sand may drain more slowly. _A physically active life may allow us to approach our true biogenetic potential for longevity._"

A BLUEPRINT FOR AGING MORE SLOWLY

Our research, and that of others, shows that people can reverse—or at the very least retard—many of the physiological declines associated with aging without turning their lives inside out and upside down. What's required is enough time to read and absorb the new information contained in this book and about 50 minutes a day to put this new knowledge into practice. Granted, we all have time pressure, but we can certainly spare an average of 50 minutes, especially when the stakes are so high. After all, an investment of a little time, a moderate boost in daily energy expenditure, and a minuscule amount of money are what it takes to improve the chances you'll live out your life less encumbered by the enfeebling symptoms of chronic degenerative disease.

Our goal in this book is twofold: First, to give you a framework for thinking about human aging, including some new facts about what's inevitable and what's not. Second, to give you the means—in the form of specific advice—by which you can slow down some aspects of the biological maturation process that's taking place in increments every single day of your life.

View this book as a combination research report, self-assessment, and action plan all trumpeting the same message:

Exercise is the key to a healthy and rewarding old age.

It's hardly a revolutionary idea that exercise is beneficial. In this day and age, who would dispute it? The idea that exercise is beneficial even for the frail elderly, however, is a departure from the norm. Our research shows without a doubt that a regular aerobics, flexibility, and strength-building exercise program, such as the one you'll encounter in this book, can have a strong positive as well as synergistic impact on the health of almost all older people. These three forms of exercise in combination are the best way we know to retard, even reverse, the progression of those inevitable—although modifiable—physiological functions we call "Biomarkers."

No, we haven't discovered the proverbial Fountain of Youth, that antidote to aging that Ponce de León and so many ancient mariners searched for in vain. Our Biomarkers are not magic elixirs, alchemic concoctions, or even easy answers to the riddle of aging. Rather, they are physiological functions—natural processes—that take place inside your body. They do, indeed, decline over time. *But the hopeful thing about them is that their decline can be arrested by beneficial lifestyle changes.* Yes, our ten Biomarkers of vitality can be controlled—and to a far greater extent than heretofore realized.

Part I

BIOINTERVENTION–
TO RETARD OR REVERSE
THE AGING PROCESS

POSTPONING YOUR ENTRY INTO THE DISABILITY ZONE

Possibly the greatest misconception people have about the process of aging is that it's synonymous with illness. It's true that chronic conditions such as heart disease are more common as we move up in years. But are such conditions a natural consequence of aging? Our research at the Human Nutrition Research Center on Aging (HNRCA) provides evidence that the link between chronic disease and aging is by no means as simple and straightforward as researchers once thought.

There are two principal factors responsible for the onset and severity of most chronic degenerative conditions—your genetic heritage, which you cannot control; and your lifestyle, which you can and should control.

The goal of our Biomarkers Program, outlined in this book, is to maintain vitality into old age. It grew out of the realization that there are specific types of exercise and eating patterns that can greatly diminish the chances that people will develop a chronic disease; or, if they're already suffering from one, could help them escape from the imprisonment of its debilitating symptoms. How? By positively altering certain key physiological functions we call "Biomarkers."

The pillars of our Biomarkers Program are three forms of exercise—strength-building and flexibility workouts aimed at your muscles; and aerobic or endurance forms of exercise aimed at your cardiovascular system. Besides quitting smoking (if you do), we believe there's no single thing that will increase vitality at any age more than exercise. *Exercise, for reasons we'll outline in this book, is the prime mover in the drive to preserve vitality.*

Granted, eating the right foods in the proper amounts is important to maintain health; but good nutrition alone will not have as appreciable an impact on most of our ten Biomarkers as exercise. Combine good nutrition with regular exercise, however, and you've got an unbeatable team. We can't think of a better antidote to fast aging.

Our Biomarkers Program is designed to
- prolong vitality by retarding or even reversing the usual biological deterioration process that people past 45 often begin to experience—such things as metabolism slowdown, glucose intolerance, and declining strength,
- postpone disability by reducing the risk for such chronic conditions as heart disease, Type II diabetes, arteriovascular disease, hypertension, and osteoporosis,
- prevent the development of a common old-age malady we call "sarcopenia."

AN AGE-OLD PROBLEM FINALLY GETS A NAME

Think for a moment about any frail elderly people you know or have known. Maybe they aren't suffering from any serious chronic illnesses such as heart disease or cancer, yet for years they've been dependent on others for care. They spend most of their day sitting, and when they bother to get out of that easy chair at all, it's a struggle. In fact, they may need assistance to accomplish just the simple feat of going from a sitting to a standing position. They walk very tentatively—probably with the aid of a walker or cane. It's as if their muscles no longer have enough power to deal with their body weight, let alone with even the most minimal activity. While their minds may still be quite nimble and alert, their bodies are diminished. They've lost so much muscle tone, their flesh seems to droop. In layman's terms, you could say their bodies have "rusted out" like an old piece of machinery nobody uses anymore. These old people have no diagnosed disease to explain their lack of oomph and vitality, yet it's clear that they've got bodies that have long since passed their prime.

We submit to you that such people have diminished to the point where they are suffering from a disease, only it's one that's never been given a name before. We're giving it a name here in the hope that it—like osteoporosis, arthritis, and the other degenerative conditions of old age—will begin getting widespread public attention.

We chose the term *sarcopenia* because "sarco" in Greek, the language from which so much medical terminology emanates, refers to flesh, the body; "penia" means "reduction in amount or need." Hence, we're describing an overall weakening of the body caused by a change in body composition in favor of fat and at the expense of muscle.

Although relatively little attention has been focused on the high-fat/low-muscle-power condition of sarcopenia, it's an extremely common condition in the elderly. Visit any nursing home and you'll see myriad examples of sarcopenia. Indeed, it's probably the most prevalent ailment.

In all fairness, sarcopenia has been overlooked for good reason. Its deleterious contribution to other old-age diseases is complex, subtle, and in many cases still little understood. Moreover, in contrast with more dramatic chronic diseases whose cluster of painful symptoms overtake a person in a matter of months or a few years, sarcopenia is a gradual wasting away of the body over the course of decades.

As they age, potential sarcopenia victims in their middle years slip into a sedentary way of living. The more they sit around, fail to exert themselves, and are waited upon by others, the greater the amount of their body's muscle mass that is replaced by fat. This insidious weakening of body structure and gradual loss of functional capacity then becomes a good excuse for continuing the pattern of immobility.

In the best of all possible worlds, of course, we'd have a coordinated national health policy to combat sarcopenia, as we have for osteoporosis (which is sometimes called "osteopenia," i.e., too little bone). We'd have a nationwide education campaign aimed at preventing sarcopenia; and we'd segregate those already suffering from it, perhaps treating them in special sarcopenia clinics where they'd receive daily therapy to reverse their condition. Our Biomarkers Program is an example of such treatment. For a finite period in residence, the frail elderly would work toward the goal of altering their body composition in favor of muscle and at the expense of fat. Eventually, when functional capacity was restored, they'd be released—with the understanding that they would continue active therapy at home. In short, they would continue getting well on their own time.

Unfortunately, we are not yet living in the best of all possible worlds as far as sarcopenia is concerned. Since sarcopenia, to date, isn't even recognized as a problem, most of the frail elderly suffering from it get no treatment whatsoever to reverse its debilitating consequences. Millions of old people in the United States remain in nursing homes suffering from this unheralded low-muscle-power infirmity. And, all the while, public policy-makers wring their hands worrying about how the nation will ever pay for the baby boom generation's expected decrepitude three decades hence.

It's important to understand that *sarcopenia is not a necessary or normal component of aging*.

THE DISABILITY ZONE AS MOTIVATOR

Sarcopenia, like such chronic conditions as cardiovascular disease, Type II (maturity-onset) diabetes, hypertension, and osteoporosis, is associated with a sedentary lifestyle, too little exertion over a long period of time. Its remedy is the converse: the type of physical activity outlined in this book—muscle-building and aerobic exercise regimens that people undertake for the rest of their health span.

Note we used the term *health span,* rather than *lifetime.* The distinction is an important one. Your health span is that time when you're functional and able to perform everyday-life tasks for yourself. You're self-reliant and capable of independent living in a natural home environment. In other words, your "functional capacity" is good. In contrast, your functional capacity is low when you can no longer lift a full garbage can and carry it to the side of the road or even drag it there . . . when that one-mile walk to the nearest bus stop or climbing two flights of stairs becomes so taxing you no longer attempt it . . . when you notice you can't stretch up and get something from the top shelf in your closet anymore.

The notion of "functional capacity" is more qualitative than quantitative, based more on observation than on narrow scientific criteria. Although comedian George Burns, America's favorite geriatric, probably never heard the term *functional capacity,* he gave his comments on the subject in his epistolary book, *Dear George: Advice and Answers from America's Leading Expert on Everything from A to B.*[1] One fan asks:

Dear George—

As one geriatric to another, maybe you can tell me— at what point does one graduate from elderly to old?

Another Geriatric

Dear Geri—

You'll know you're old when everything hurts, and what doesn't hurt, doesn't work; when you feel like the night after and you haven't been anywhere; when you get winded playing chess; when your favorite part of the newspaper is "25 Years Ago Today"; when you're still chasing women but can't remember why; when you stoop to tie your shoelaces and ask yourself, "What else can I do while I'm down here?"; when everybody goes to your birthday party and stands around the cake just to get warm.

(These things really happen when you get old. I know, because that's what my father keeps telling me.)

George Burns is a nonagenarian with good functional capacity despite his chronological age. He was born on January 20, 1896, and unlike many of his chronological peers whose get-up-and-go long since got up and went, he's still a contender. As of this writing, he's dispensing witty advice, similar to what you just read, both in books and on radio and TV talk shows.

In contrast with George, older people with a low functional capacity who can no longer care for themselves can be said to have entered the "Disability Zone"—that period, usually at the end of a person's life, when he or she becomes increasingly dependent on others.

The average man-in-the-street perceives old age as a sojourn in the Disability Zone. Stop people at random on any street corner, and here's a representative sampling of what you'll hear about that dreaded Disability Zone, although no one will use the term:

"Old is when physical impairments rule your life."
Or . . .
"It's when you've allowed your body's weaknesses to take control of your mind and spirit."
Or . . .
"You know you're old when you sit and watch as other people perform the everyday rituals of life for you."

Our research shows that the states of decrepitude described in these statements can be altered through the Biomarkers Program described in this book. You could think of our Biomarkers Program as the means to slow and flatten out your descent toward the Disability Zone. As the schematic (figure 1-1) shows, making the transition to a healthier, more energetic lifestyle via our Biomarkers Program at as early an age as possible promises to add years of vitality to your life—and postpone for decades your entry into that unfortunate Disability Zone.

Sarcopenia is one of the main ailments that forces people into the Disability Zone. Thus, we feel if sarcopenia can be avoided, a person stands a very good chance of bypassing that dreaded Disability Zone altogether. Although we have no longitudinal studies to prove it yet, we believe that middle-aged people who remain faithful adherents to an active Biomarkers lifestyle throughout their 50s, 60s, and 70s could find themselves in the fortunate position of having their health span approach their life span.

Some of you may be wondering: Can a transition to a more active lifestyle via the Biomarkers Program increase my longevity?

Based on the current research, we can state unequivocally that people who engage in mild but regular exercise have a longer life expectancy than those who are sedentary.[2] What we don't have as yet are statistics on the life expectancy of sedentary people who begin an exercise program in their 50s or 60s. We cannot say with certainty that they will live longer than if they stayed in that rocking chair. We suspect they will, but a definitive answer must

Health Zone

Biointervention

Making the
Transition to
a More Active
Lifestyle
via the
Biomarkers
Program

Decline in those inevitable
but dynamic Biomarkers

Inactive Lifestyle — little or no formal
exercise

Active Lifestyle — incorporating aerobic and
strength-building exercise and proper diet

Postponing Entry into the Disability Zone —
extra years of health you could gain through Biointervention

Age in
Years:

45   50   55   60   65   70   75   80   85   90

Disability Zone

await the conclusion of longitudinal studies encompassing the sweep of a lifetime.

We can say this: If you change your lifestyle and undertake the Biomarkers Program, *the quality of your remaining years will be greatly enhanced.* You'll avoid that Disability Zone much longer than you would have by remaining in that easy chair.

AS MANY AGING SCHEDULES AS THERE ARE PEOPLE

Some gerontologists have likened the process of biological aging to a long, slow tide that moves upon us so gradually it may take a long time before we're aware of its presence. Certainly aging is not something that happens suddenly. It moves in on us in increments over the whole middle part of our lives. As Butch Cassidy remarks ingenuously to his sidekick, the Sundance Kid, in the film bearing their names: "Every day you get older. That's a law." What's not an immutable law, though, is the pace at which you get older. That varies enormously from one individual to the next.

The beat of your biological drummer is different from ours and the next person's. Indeed, older people are more *unlike* their peers than younger folk. By a long shot. Decades of studying both normal and abnormal aging—that is, premature aging due to illness—have convinced investigators beyond a doubt that *the older people become, the less like each other they become.* Some people are weak and withered and wrinkled at age 60. Others, at age 75, are energetic, eager for new adventures, and look and feel far younger than their chronological age.

This fact has certainly been brought home to us at the HNRCA, where we've recruited thousands of volunteers for research studies over the last five years. Every week we interview older people whose chronological age, appearance, and medical test results surprise us. We'd like you to meet three senior citizens, all people who volunteered for our research studies. In terms of their aging timetables, they represent different ends of the spectrum:

Frank was 66 when he came into our center for a screening interview. He was a distinguished-looking man, and his comportment indicated he knew it. Frank characterized himself as "health-conscious," but as he talked a different picture emerged.

Frank had a whole litany of minor ailments and annoyances,

but he'd only gotten treatment for those visible signs of aging that showed when he looked in a mirror or that he knew other people would notice. Although he certainly understood the connection between smoking and cardiovascular disease, he felt he was somehow immune. His father had lived to a ripe old age, why shouldn't he? Frank is like a lot of people in our society. He sees health through the prism of appearance and vanity. He has little concept of what real fitness or physical well-being are all about.

In all fairness, Frank abhorred the idea of losing his youthful looks for a very practical reason. He equated the external signs of health with economic survival. Frank had spent his career in an industry where youth and attractiveness are highly prized, and aging in an ungraceful fashion could be the unspoken grounds for dismissal.

For most of his career, Frank was a salesman for several leading national magazines, making the daily rounds of advertising agencies and executive suites soliciting pages of advertising. Even though the job was high-pressure, the glamour and chance to live well on a lavish expense account appealed to him. Sometimes he found it hard to believe he was actually being paid a good salary to enjoy life.

But by age 49 Frank had grown tired of all the glad-handing and free spending and decided he wanted a different type of job. He'd always had a flair for words, so he decided to use his advertising agency contacts to get a copywriting job. A string of interviews that didn't pan out and a long look in the mirror convinced him that his age was showing more than he'd thought. He resolved to do something about it.

Frank subscribed to several health magazines and bought a battery of vitamin and mineral pills. He went on a crash diet and did muscle-toning exercises in his bedroom until he lost the beefiness in his face and girdle of body fat he'd started to develop around his middle. Even though his eyesight was poor, he decided he could no longer afford to be seen in eyeglasses, especially bifocals, a sure sign of advancing age. He started wearing contact lenses and later switched to bifocal contact lenses, even though they gave him constant trouble and he couldn't see very well. The final touch was an expensive hairpiece to cover the signs of his receding hairline.

Frank landed that copywriting job and was in that line of work when we met him. His true health and fitness were another story.

When we took a medical history, Frank admitted, with some embarrassment, that he'd had a little plastic surgery once; and he'd also tried a number of skin-care elixirs that made claims of eliminating wrinkles and firming the skin. Frank's hearing was impaired, but he'd done nothing about it because he refused to wear any ear device, no matter how small. Frank was startled when he discovered he'd already shrunk a half inch from his peak height of 6 feet. Although he weighed only 10 pounds more than he had in college and he looked relatively trim, his body had gained fat and lost considerable muscle. He now had a lean–body mass/body-fat ratio of 4:1. The body fat gain was concentrated in his belly. At 20 he'd had a 36-inch chest and 33-inch waist. Our tape measure showed he now had a 37-inch waist.

Frank's muscular strength was below average. He'd never lifted a weight in his life, nor did he see much sense in the stretching exercises his wife did in front of the television set in their bedroom every morning. He adamantly refused to join her, and the flexibility and range of motion in his joints were the worse for it.

Frank was an off-again, on-again jogger and a once-a-week tennis player. His cardiovascular endurance on a treadmill showed it. He was below the norm for his age. Likewise, his blood pressure was high for a man of his age and race.

We asked Frank to describe a typical day.

Frank explained that he and his wife live in a condominium community geared toward people aged 50 and over. The grounds are maintained beautifully by professional landscapers, and maid service is even provided for an extra fee, so he feels it's the ideal place to work at home, his current situation. Despite the fact that he's a free-lancer, he maintains 9-to-5 hours. Breakfast is three cups of black coffee and several cigarettes, which he consumes while scanning the morning newspaper. His wife is off to work by 8:30, about the time he makes it over to his computer to start his workday writing advertising copy. Work consists of reading, writing, and talking on the telephone, punctuated by occasional visits to clients' offices or a business luncheon.

His wife gets home around six, and an evening's worth of congeniality and good food commences. The couples within their circle of friends pride themselves on their culinary skills and go all out preparing meals for each other. Frank and his wife dine out four or five nights a week, either at friends' homes or in restaurants. He gets to bed around 1 or 2 A.M. but has a problem with insomnia. He never lies in bed and worries about it, though. Instead, he reads a book or settles in front of the TV set with his cigarettes and cognac to watch a late-night movie.

While awaiting further evaluation to become a participant in one of our studies, Frank suffered a stroke that left him partially paralyzed on the right side of his body. At our last contact Frank was at home, undergoing daily rehabilitative therapy.

Frank's story is one of premature disability brought on by a misconception of what constitutes good health habits. We tell it by way of contrast with an older couple we screened around the same time.

George and Elaine McGrath live in rural Vermont on a nonworking farm they bought 50 years ago when they were young parents. It was the ideal country retreat, an unpretentious, semi-rundown place where they could turn their five children loose to roam in nearby fields and woods during summers and on weekends.

Years later, when their children were grown, they questioned whether they wanted to continue being responsible for the ramshackle, nineteenth-century house and outbuildings, not to mention the ten acres of meadows that went with it. They put the property up for sale. But after two months and no takers, George admitted he missed the challenge of caring for the place and felt cooped up on weekends in their suburban ranch home in the Boston suburbs. They decided to keep "the farm," as they call it. Later, when they retired, George talked Elaine into moving there permanently, and neither of them have regretted it.

George, a former accountant, and Elaine, ex-administrator of a social services agency, have never made a conscious effort to exercise, nor did their white-collar occupations demand any strenuous activity. The upkeep of the farm was—and still is—the only thing that forces them to use their muscles to any great extent.

Years ago the whole family renovated several of the outbuildings. Since their retirement, the two of them, with minimal outside help, have made further major improvements to the property. They've remodeled the main house and, as recreation for their visiting grandchildren, added a patio and swimming pool. The adjoining heated whirlpool bath is for their pleasure and relaxation. It's a godsend after a hard day of labor in their quarter-acre fruit and vegetable garden or when used to ease their sore muscles after a vigorous wood-chopping session.

The couple assured us they're not lonely, nor are they alone, despite the fact that they live on the crest of a hill two miles outside the nearest town. Their housemates are six cats, a big shaggy Scottish sheepdog who seems to produce puppies on a fixed schedule, and George's loyal Irish setter. The dogs accompany them on long daily walks, sometimes to do errands in the town nestled in the valley below them. Since they're snowed in for several weeks each winter, making vehicular travel perilous, foot power is often the only way to restock the refrigerator with the bare essentials. Besides, they say they've gotten hooked on their long walks, not just as a practical necessity, but as a form of enjoyment since the rolling landscape in their vicinity is very beautiful.

Their health?

George had open-heart surgery five years ago to replace a malfunctioning cardiac valve, which had caused him to suffer fainting spells. After an uneventful recovery, George said he felt better than ever—and greatly relieved that he no longer had to worry about fainting at unexpected times. Except for various allergies and occasional attacks of bursitis, he has no major complaints. He said it took him much longer to do things now than when he was younger, but the more leisurely pace was nice. It gave him time to savor each project.

Elaine was also in good health and spirits. Years earlier, as a frazzled working mother, her crucible had been severe, incapacitating backaches. But for the past 30 years she'd done a series of stretching and back-strengthening exercises each morning that offered relief. Now, with no time pressure and stress in her life, the backaches were gone. She had arthritis in her hands, but she found she could reduce the pain by playing technically demanding music on the piano or the organ of the local church, where she was the choir director. While Elaine had lots of facial wrinkles from the many years she'd spent relishing the sunny outdoor life, she never mentioned them, nor the fact that her posture was somewhat stooped and she'd shrunk almost two inches from her peak height of 5 feet 6 inches.

Have you guessed their ages yet?

At the time of our first meeting, George was 81 and Elaine was 77. Clearly, neither of these two vital older people was anywhere near entry into the Disability Zone. Their active lifestyle and better-than-average results on various age-adjusted Biomarker tests were proof enough of that.

Compare George and Elaine's situation with that of Frank. At 66 Frank has already entered the Disability Zone, perhaps to emerge from it, perhaps not.

DISEASE-FREE AGING IS PARTLY WITHIN YOUR CONTROL

As these examples show, people can, indeed, deflect their descent into the Disability Zone by controlling the key physical aspects of aging that are, to a great extent, *within human control— your control!*

We know that the decline in our ten Biomarkers of vitality can be halted, or at least slowed down considerably, through our exercise and diet program. All ten of these Biomarkers are amenable to intervention via our program—what we call "Biointervention." We have studies to prove they're amenable to change even if you're already 75 years old and feel overburdened with aches and pains and have long since accepted the idea that an energetic lifestyle is over for you.

No, you're never too old to exercise. To the contrary, you're too old *not* to exercise. That's certainly the lesson that 12 men, aged 60 to 72, learned when they participated in a landmark 12-week muscle-building study we undertook not long ago. If you've got doubts about our claims, consider the results achieved by two of our participants:

Manuel S., a retired banker, was 70 years old when he volunteered for our strength-training study. During the screening interview, he admitted rather sheepishly that the last time his three-year-old granddaughter visited, he was horrified to discover he could barely lift the little girl off the ground. A year earlier he'd been able to pick the child up and whirl her around in the air without even thinking about it. While the child had certainly gotten heavier in the interim, she hadn't grown enough to account for his drastic drop-off in strength. Upon reflection, Manny realized his body was becoming weaker. He could feel it in his arms, back, and legs.

Here's what 12 weeks of strength training did for Manny:

• When the study started, Manny could lift 25 pounds. When it ended, he was able to lift 75 pounds. Picking up his granddaughter became less of an effort once again.

- The size of his leg muscles increased by 17 percent.
- His aerobic capacity increased markedly to the point where he could chase his granddaughter around the backyard.
- Over the course of the study, while he was gaining muscle, he simultaneously lost almost 15 pounds of useless body fat.

Like all of our study participants, Manny was so proud of these startling physical achievements that he didn't want to lose what he'd gained. Seven years earlier, at age 63, he'd given up playing soccer. He loved the game but found it was too difficult to run up and down the field. Our study proved to Manny that his body could be rejuvenated. He continued to engage in regular aerobic and strength-building workouts after the program ended. Eventually, when he felt he was back in peak shape, he rejoined his soccer club. Today he claims he's playing better than he did when he was 20 years younger. He'd better be. His teammates are all 30 to 40 years his junior!

Arthur F. was 62 years old and still working when he participated in our study. For years he'd had a physically demanding job as a loading dock worker. At age 50 he was promoted to foreman, which required much less exertion. He, too, was jolted into volunteering for our study by a humiliating incident. One day he'd tried to help one of his subordinates lift a heavy crate of scrap metal, but even with help the chore overwhelmed him. Of course, his men noticed and let him know about it. The barbs were more than Art could bear. "We understand how it is, Arty. . . ." "At your age, the only thing you should be lifting is a beer can. . . ." "Step aside and let the younger guys do it. . . ."

That night Art took a long private gander in a full-length mirror. What he saw wasn't pretty—an overweight, out-of-shape, over-the-hill guy with a large belly hanging out over his belt. The way he felt wasn't wonderful, either. Among other things, he suffered from mild but chronic lower back pain.

When Art began the study, he had an abnormally high circulating insulin that indicated there could be diabetes in his future if he didn't do something fast to reverse the trend.

Here's what happened to Art after 12 weeks of strength training:

- His strength increased from being able to lift 50 pounds to lifting almost 150.

- He lowered his fasting insulin level to normal, which we ascribe to the fact that his reactivated muscles became more sensitive to the hormone insulin and, thus, started to withdraw more of it from his blood.

Art got so hooked on weight lifting that he, too, continued to do it after the study ended. As a consequence, he strengthened his lower back muscles, and to his great joy, his chronic backache disappeared. He also discovered that the more muscle he built, the more calories he seemed to expend every day even without trying. He didn't change the amount of food he ate, but he did consciously cut down on the fat in his diet. Over time he was able to lose 25 pounds and reduce his stomach circumference by 7 inches.

From Art's point of view, the best bonus of all was being able to keep up with the younger guys at the loading dock, much to their amazement and his amusement.

We feature only two of our study participants here because it would take too long to relate the stories of all 12 men. We can give you their collective results, however. After only three months of working out . . .

- the muscle strength of all 12 men increased two- to threefold.
- their muscle mass grew by 10 to 15 percent.

When you learn more about how the all-important Biomarker of Strength—the first on our list—influences the others, you'll understand the long-term health implications of these results. Suffice it to say here, if these men continue to work out and increase their muscle strength and size without increasing the amount of calories they eat daily, their exercise efforts will yield the additional benefit of replacing the amount of harmful excess fat on their bodies with muscle.

The moral in all this is simple: If you're in good enough health to hold up this book—which weighs several pounds—and your mind is nimble enough to absorb our message, you can still stage a physical comeback. We don't care how old you are.

ENGAGING YOUR MIND AS WELL AS YOUR BODY

How is our program different from all the other fitness-over-50 books?

This book transcends the other approaches to extending the human "health span" for this reason: Unlike those other books and videos, ours is grounded in research-based theory supporting the notion that exercise and a proper diet are good for you. Our program places heavy emphasis on what we're learning daily in our laboratories and through our clinical research studies using *real older people* as subjects.

To be sure, the solution to biological degeneration is still beyond our reach. No, we scientists at HNRCA have not yet cracked the mystery that has perplexed mankind since the beginning: We still don't know why it's an immutable law that every human being deteriorates over time and eventually dies. But at least we feel we've arrived at the point where we're now asking the right questions. While it's true that formulating answers to these questions involves a lot more research, much of it is already under way and ongoing.

In other words, in this book we'll not only be advising you what to do, we'll be explaining why we're asking you to do it.

In the next chapter, we'll introduce each Biomarker and show how exercise, sometimes in conjunction with a dietary change, can go a long way toward preserving health even into old age. We'll also be giving you self-diagnostic tools that will enable you to decide for yourself how far your body, regardless of your chronological age, has already proceeded along the normal aging continuum. How much of your body mass has accumulated fat at the expense of beneficial muscle tissue? What does the distribution of fat on your body say about your risk for developing certain age-associated ailments?

Part Two of the book focuses on two important self-tests. They'll enable you to find out if your aerobic capacity and muscular strength place you in the category of low fitness, average fitness, or way above the norm for a person of your age and sex.

In Part Three we put you to work. Depending on your fitness category, which you will know from taking the tests in Part Two, we ask you to follow a detailed exercise program we call a "BioAction Plan." Think of this structured program as a transition to a more exuberant lifestyle. Once you've completed our formal BioAction Plan, you'll be able to follow easier—and less time-consuming—exercise guidelines from then on.

We also offer a special section for those older competitive athletes, amateur though they may be, who take exercise very seriously. Drawing on the latest research, we'll erase misconceptions about training and offer information designed to improve performance. Our goal is to help readers in this very special category participate in their sport for many more satisfying years ahead, for we feel chronological age should never be held up as the barrier to athletic endeavor.

Part Four is devoted to nutrition and as much as we currently know about its role in the aging process.

TAKING THE VITALITY VOW

While our Biomarkers Program is researched-based, don't let the notion of medical research scare you. Our program is not hard to understand. Certainly you'll have no trouble comprehending its underlying concepts. Nor is its execution beyond even the most sedentary person's ability, for the program is graduated to take all levels of health and fitness into account. However, we do think you may be challenged by the constancy and commitment our program requires.

Biomarkers is not a program for the unregenerate spectator, fence sitter, benchwarmer, or couch potato. *It's a program for participants.* Doing the Biomarkers Program justice will require effort on your part almost every day of your life. Our exercise and diet guidelines will stand you in good stead, healthwise, for the rest of your life—*provided you follow them for the rest of your life.*

One whole chapter in this book is devoted to motivation. Make sure you read it carefully. As strong-willed as you may think you are, you could find, as we have, that aching bodies and unexpected emotions have a way of impeding logic. Your head may tell you that exercise and eating right are sensible things to do, but the rest of your body may act as a saboteur. We give you the weapons to ward off such recalcitrance. We've designed our Biomarkers Program with the goal of capitalizing, as much as possible, on the more sociable, fun aspects of exercise.

In the early 1980s, in conjunction with the Massachusetts Department of Elder Affairs, we started a walking program called "Keep Moving, Fitness After Fifty." The program was built on the pyramid principle. We trained a select group of older people to be the leaders of walking clubs in their community. They, in turn, recruited participants from among friends and friends of friends. The program continues to this day with more than 7,000 registered walkers (average age: 65), and the numbers keep multiplying.

We think the program continues to gain popularity because walking, of all the aerobic sports, is such a congenial way to exercise. It's also why the aerobic component of our Biomarkers Program is centered around walking. It's fun, it puts less stress on aging joints, and it's a way to make new friends while you're improving your sense of health and well-being.

The crucial strength-building component of our program can also be done in the company of friends—and we think it should be. In fact, we urge everyone who undertakes our Biomarkers Program to do it with at least one other person and, preferably, with an amiable, heterogeneous group of friends.

A PRESCRIPTION FOR LENGTHENING YOUR HEALTH SPAN

The message of our Biomarkers Program is a straightforward one:

You can adopt a pattern of activity and eating that maximizes your ability to age much more slowly.

Heredity isn't everything, after all. It's certainly possible for a well-maintained Volkswagen bug to last longer than an abused Mercedes, isn't it? By the same token, an active and independent old age is within the reach of any middle-aged person who is willing to get out of that easy chair.

No, it's seldom too late to inject more pep and vitality into your life, not if you're determined enough—and have the willpower to make the effort. We'll tell you what to do, but *you have to do it!*

The idea you should keep in the forefront of your consciousness like a beacon is that *your health largely depends on you.* You can fill your own prescription for successful aging. *No matter what your age, it's not too late to turn over a new leaf and start building strength, adding to your endurance, and eating right.*

CHAPTER
2

THE TEN BIOMARKERS OF VITALITY

———

Aging is one of those vague terms that all of us use constantly but even many medical doctors can't define with precision. We think we know what aging looks and feels like. We can point to all manner of friends and relatives who are "old." But a specific definition eludes us.

Traditionally, the scientific community has studied aging by focusing on declining biological functions that can be used to assess how much a person's body has proceeded along a normal aging continuum.* This continuum charts the biological milestones the human body undergoes over the course of a lifetime.

---

* A normal aging continuum specifically *excludes* chronic, degenerative illnesses because there's nothing normal or natural about them. They are disease conditions even if they are related to aging and found, more commonly, in an elderly population.

At the left end is birth, followed by the growth transformations of childhood and adolescence. At the far right end is old age and death.

Leading biologists have also dealt with the phenomenon of aging by trying to isolate those declining biologic processes that render older people more susceptible to potentially fatal diseases. Alex Comfort, the English gerontologist, has defined aging as "an increased liability to die."[1] Maynard Smith, another Englishman, defines aging processes as "those which render individuals more susceptible as they grow older to the various factors, intrinsic or extrinsic, which may cause death."[2]

At the USDA Human Nutrition Research Center on Aging (HNRCA) at Tufts, we realized early on that we needed a better way to think about and measure aging.

We discarded the chronological approach straight away because it wasn't accurate when we got down to actual cases. Biologically, older people are just too different from one another to measure aging in years. While there is a certain inevitability about growing older, it cannot be marked off by the steady rhythmic beat of a metronome or ticking of a clock; nor can it be expressed as a downward line slipping off a piece of graph paper.

Defining aging from the perspective of increasing risk of death didn't strike us as fruitful either. It's like viewing a glass as half empty instead of half full.

After debating the merits of various positions, we finally decided on what we at the HNRCA feel is the most constructive vantage point. Instead of looking at aging from the negative (what causes death), we've been concentrating on the positive (what maintains health). That is, *our focus is not on postponing death, but on maintaining health for the longest possible period of time.*

We want people to avoid the Disability Zone for that period, typically found at the end of life, which is characterized by invalidism, dependency, frailty, and pain. In short, *we want to find ways for older people to* **maintain vitality** *for the longest possible period of time.*

PEERING THROUGH A NEW SET OF LENSES

Once we'd settled on this vantage point, we needed to identify specific biological predictors of health. At this point, we found ourselves resurrecting a concept that's been floating around the medical research community for years. It's the term that graces the cover of this book.

The word *biomarker* is often used to refer to biological markers of age. The word has, at one time or another, been attached to just about every conceivable aspect of aging from the graying of a person's hair and the deterioration in a person's hearing ability to the blurring of vision through cataracts. If we were going to use the term *biomarkers* at all, we immediately realized we needed a much tighter definition than the traditional one.

Right off, we deleted from our long list of biomarkers the changes in appearances that age brings about. As medical researchers, we're not focusing on the cosmetic aspects of aging, those visible signs of decline—sagging skin, age spots, receding hairlines, more pronounced facial features, and the like—that cause people so much anguish and create a huge market for beauty products companies.

Even without the purely cosmetic indicators, we were still left with a rather lengthy list of biomarkers. Many of those remaining were internal, physiological functions that most people don't know much about and generally overlook when they think about growing old. They're functions that decline, to be sure, but at widely divergent rates in different people.

Given our decision to keep our sights trained on the positive side of the aging equation, even this list had to be radically culled. We did this by excluding all those physiological biomarkers that cannot be altered for the better by changes in a person's lifestyle. Examples are the deterioration in our senses of taste and smell as well as the other senses; our slower reaction time and slower speed at which nerve impulses are carried through the body; and the decrease in our vital capacity (the lungs' maximum capacity for holding air).

Sad to say, there's little any of us, living in 1990, can do to halt certain functions' inexorable decline given the present state of our knowledge. This is not to say that medical researchers won't discover ways to influence these functions tomorrow, or next month, or two years from now. We grant that a sequel to this book, written five years down the road, could undoubtedly feature a much longer list of modifiable biomarkers than we've managed to assemble here. We are especially hopeful it will feature as a biomarker the body's immune function, which we tried to include on our list but couldn't yet for lack of enough definitive data.

As the box below indicates, there are ten entries on our official Biomarkers list. We settled on these ten for two reasons: (1) they're critical biological functions that influence vitality; and (2) we know how to revive these functions, even in old people.

To paraphrase a remark by Leroy "Satchel" Paige, the famous baseball pitcher, our Biomarkers are those things that tell you how old you would be "if you didn't know how old you was." Paige never knew his exact chronological age because he was born before good records were kept. He used to tell people to use his performance on the pitcher's mound to estimate his age. But that was almost impossible, because Paige's long heyday as a leading pitcher stretched from the 1920s to 1950s.

Just like Satchel Paige, we'd like you to forget your chronological age from now on. Your age in years, after all, has little to do with how old you are biologically and how old you feel. We want you to *think about your body only in terms of the following ten Biomarkers.*

---

10 BIOMARKERS OF VITALITY THAT YOU CAN ALTER

Biomarker 1: Your Muscle Mass

Biomarker 2: Your Strength

Biomarker 3: Your Basal Metabolic Rate (BMR)

Biomarker 4: Your Body Fat Percentage

Biomarker 5: Your Aerobic Capacity

Biomarker 6: Your Body's Blood-Sugar Tolerance

Biomarker 7: Your Cholesterol/HDL Ratio

Biomarker 8: Your Blood Pressure

Biomarker 9: Your Bone Density

Biomarker 10: Your Body's Ability to Regulate Its Internal Temperature

---

THE DECISIVE TETRAD

We put the first four Biomarkers at the head of the list for a reason. They're the most closely interrelated. In addition, we view them as the primary catalysts for preventing sarcopenia. Here's why:

In middle age most people find themselves fighting the battle of the bulge—not too successfully, we might add. Compared with the citizens of other major nations around the world, Americans are too heavy. More specifically, *we carry around too much body fat and too little muscle,* a condition that hastens sarcopenia.

We'd like to give you a new slant on your girth problem: If you're the average middle-aged person, your problem is not excess weight so much as it is *excess body fat coupled with too little muscle.* If you want to avoid this decreasing muscle/increasing fat state of affairs, it's imperative that you look beyond the simplistic notion of losing weight. What we'll be asking you to do throughout this book is focus on the far more vital issue of body fat versus muscle—to *concentrate on building muscle at the expense of fat.*

The crux of the matter is a concept called *body composition.* Scientists divide the components of the body into two broad categories: *body-fat mass* and *lean-body mass.* The former is just what the term implies—fat, known by the technical term, *adipose tissue.* The latter is everything that is *not* fat—bone and other vital organ tissue, your central nervous system, and so on, but mostly it consists of muscle.

Here's why simply losing weight is the wrong goal:

Body fat is metabolically inactive. It's energy *storage* tissue. Millennia ago, when the earth was populated by hunters and gatherers, body fat served a life-and-death purpose. During periods of famine and starvation, body fat stores were a kind of rainy-day calorie bank.

In the developed countries of the world today, however, the situation is quite different. Most people aren't starving, and their excess body fat, instead of solving a problem, is often the source of medical problems. It's been implicated in a broad range of chronic and potentially lethal diseases including diabetes, hypertension, heart disease—and, now, sarcopenia.

The body's biologically active tissue consists of muscle, bone, and vital organs (lean-body mass). This tissue, especially muscle, needs a far greater amount of caloric fuel to maintain itself compared to passive adipose tissue. Thus, *people with a high ratio of muscle to fat on their frame have a higher metabolism and a higher caloric need, and they don't have to worry as much about how much they're eating or about gaining weight. Conversely, because fewer calories are needed to maintain inactive fat tissue, obese people have a lower metabo-*

lism and a harder time losing weight no matter how little they eat. In addition, people who are accumulating fat are simultaneously losing muscular strength because the muscle component of lean-body mass is lost at a more rapid rate than any of the other components.

Muscle, to a far greater extent than most people realize, is responsible for the vitality of your whole physiological apparatus. It's why muscle mass and strength are our primary Biomarkers— and why we believe that building muscle in the elderly is the key to their rejuvenation.

A strong, toned musculature makes all sorts of wonderful contributions to your overall well-being. A high ratio of muscle to fat on the body . . .

- causes the metabolism to rise, meaning you can more easily burn body fat and alter your body composition even further in favor of beneficial muscle tissue.
- increases your aerobic capacity—and the health of your whole cardiovascular system—because you have more working muscles consuming oxygen.
- triggers muscle to use more insulin, thus greatly reducing the chances you'll ever develop diabetes.
- helps maintain higher levels of the beneficial HDL-cholesterol in your blood.

In the remainder of this chapter, we'll introduce you to each Biomarker. In many instances we'll also be giving you self-analysis tools, enabling you to judge the status of a Biomarker in your body. Finally, we'll explain how our exercise program will retard—and in some cases reverse—any decline you may have experienced in that Biomarker.

BIOMARKER 1: YOUR MUSCLE MASS

Long-term studies bear out the fact that the average person's lean-body mass declines with age. These studies show that Americans, as they move from young adulthood into middle age, tend to lose about 6.6 pounds (3 kilograms) of lean-body mass each decade of life. _The rate of loss accelerates after age 45_—which is why our Biomarkers Program targets people in this 45+ age group.

Two things are responsible for how much muscle we have: The first involves how much we use our muscles. The second is the level of tissue-maintaining anabolic hormones circulating in our blood.

How much a muscle is used is partially responsible for its size and lifting capability. A muscle used frequently is a muscle that will maintain the *status quo*. A muscle that is not only used frequently but is pushed to the limits of its capacity via our Biomarkers Program will grow and gain strength, *even in elderly people.*

The second muscle-size factor is the amount of anabolic hormones in the bloodstream. Anabolic hormones increase the synthesis of protein. The most potent of these is testosterone. Because men have much more testosterone in their bodies than women, they also have more muscle. The time of the most rapid muscle growth in men is during their teenage years, which is also the period when testosterone levels are surging. The positive effects of this class of hormones is the reason why so many strength and power athletes—weight lifters, jumpers, discus throwers, and shot putters, not to mention body builders—are tempted by illegal injections of anabolic steroids, which are nothing more than a synthetic version of testosterone.

Recently, a report from the Medical College of Wisconsin caused quite a stir because it suggested that growth hormone injections in older men increased their lean-body mass, skin thickness, and density of their lower spine and decreased fat mass.[3] There are a few things that we would like to point out about this study for those of you who now think the solution to the riddle of aging is to rush out and get growth hormone injections.

First of all, the growth hormone was given only to men—not to women. They were men who already had low growth hormone levels; such a sample represents only about 30 percent of the population of people over 60 years old. Second, growth hormone is extremely expensive. For a year of growth hormone injections, you'd pay as much as $20,000. Clearly this therapy, even if it were a proven one, is well out of the financial reach of most older people.

While the Medical College of Wisconsin investigators reported increases in lean-body mass, they did not distinguish among types of lean-body tissue. They did not measure the amount of muscle in each man's body, for instance. The lean-body mass increase could well have been in such tissue as the liver and other organs. This type of change is not the beneficial muscle and strength increase we're describing here.

One study finding, in fact, is less than promising for the elderly. The study reported that fasting blood-glucose levels and blood pressure rose significantly after the growth hormone injections. Both of these changes are important considerations because many older people are already struggling to combat the potentially dangerous conditions of hypertension and Type II diabetes.

While the results of this study may appear promising for certain older people, that universe is relatively small. In general, we think there are just too many questions yet to be answered before we can recommend the widespread use of growth hormone.

BIOMARKER 2: YOUR STRENGTH

Building muscle tissue—in short, regaining and maintaining strength—is one of the principal goals of our program because a healthy musculature has broad implications for those other Biomarkers of vitality on our list. To give you more insight into these interrelationships, bear with us for a moment as we go into some detail about your body's musculature.

You've probably heard the term *skeletal muscles*. The skeletal reference simply points up the fact that muscles are attached to bones. It's your muscles—with the help of nerves—that cause your bones to move. Here's how it works:

Motor nerves connect your skeletal muscles to your central nervous system. These nerves carry messages back and forth much like a telephone wire. A given set of nerves and muscles are known as a "motor unit."

Here's the problem: As we age, we lose whole motor units. In cross-sectional studies, it's been estimated that over the 40-year span between age 30 and 70, people experience a 20 percent decrease in the number of motor units in their thigh, for example. Similar decreases are sustained in both large and small muscle groups all over our body.

There's another aspect to this muscle-atrophying process. Muscles are made up of two kinds of fibers—the fast-twitch and slow-twitch. Slow-twitch muscle fibers are necessary for posture and most low-intensity movement. The fast-twitch muscle fibers, in contrast, are the kind we mobilize when we strain to lift heavy objects or to do high-intensity, sprint-type exercise. It's these fast-twitch muscle fibers that decline with age.

Swedish investigator Lars Larsson performed muscle biopsies on a large sample of people, from teenaged boys and girls to people in their 80s and 90s. He noticed that the number of fast-twitch fibers falls with age. He speculates that part of the decline is attributable to mature adults' increasingly sedentary lifestyle.[4] In short, as their activity level goes progressively down, their fast-twitch muscles atrophy. Since fast-twitch fibers are called into action only during high-exertion exercise, there appears to be a lot of truth to the old saw, "Use them or lose them."

On the other hand, Larsson also noticed that even in older people who remained physically active, there was an age-related fall in the percentage of fast-twitch fibers. He thinks this drop-off may begin as early as age 20, which would explain why world-class sprinters peak in their late teens and early 20s. Since a high percentage of fast-twitch fibers is essential to an elite sprinter, even the most subtle change in fiber composition can make a big difference in a sport in which success is measured in hundredths of a second.

Other researchers have also formed hypotheses about the cause of this fast-twitch-fiber atrophy, which has the effect of a very gradual muscular dystrophy. Some suggest that the age-related changes that our central nervous system undergoes— which appear to be inescapable and nonreversible—cause a loss in motor nerves, mainly the motor nerves connected to fast-twitch muscles.

But no matter what the reason for it, the end result of this long deterioration process is the same—more slow-twitch and fewer fast-twitch muscles and fewer motor nerves in a person's body. This translates into slower, more measured body movement and partially explains age-related loss in strength.

Like everything else in your body, your muscles are made up of individual cells. No study has ever shown that it is possible to increase the total number of muscle cells that you have, although ours at Tufts show that vigorous exercise can cause a phenomenon called "muscle hypertrophy," meaning individual muscle cells grow larger. What studies have indicated is that from age 20 to about 70, we lose almost 30 percent of our total number of muscle cells. Not only that, but with age, the muscle cells that remain start to atrophy. Each individual cell gets smaller. As a consequence, the muscles can't contract with as much force. End result: _decreased muscular strength_.

You could think of your muscle mass and strength as the lead dominoes in a lineup of Biomarkers. As these first two dominoes start to topple, so do all the others in turn—albeit ever so slowly.

Gradual muscle loss is the catalyst for a number of other age-related changes in your body—which is precisely why our program places so much emphasis on rebuilding and maintaining strength as the best way to postpone senescence. These adverse changes are . . .

- a slowdown in your metabolism
- a steady increase in body fat
- a declining aerobic capacity
- a reduced blood-sugar tolerance
- a continuing loss in bone density

The Impact of Exercise on Muscle Strength

Today, thanks to several studies headed up by colleagues of ours at Tufts, we can make a bold assertion: A decline in muscle strength and size is not inevitable.

Walter Frontera, a physician trained in rehabilitation medicine, came to our laboratory in 1983 to pursue a Ph.D. degree in exercise physiology. For his thesis project, he decided to examine the effects of weight lifting on the size and strength of elderly people's muscles. Walter wanted to test two long-held beliefs about aging. The first is that the older you get, the less strength you're able to gain in response to weight lifting. The second is that any increase in strength you do gain results from "learning" and *not* from muscle hypertrophy—that is, muscles growing larger.[5] The conventional thinking is that age somehow decreases the ability of the muscles to get bigger in response to strength training.

Walter's landmark research turned these two ideas on their ear. Nonetheless, both are still firmly believed by most lay-people, and there are still references to them in the scientific literature.

Where did earlier researchers go wrong? In many strength-building studies, the investigators did not push their older subjects hard enough. The researchers assumed their older subjects couldn't withstand anything beyond very low intensity training. Usually this meant 30 to 40 percent of their maximal lifting capacity.

But exercise physiologists know that to improve strength in young people to any worthwhile degree, they must lift between 60 and 100 percent of their maximal capacity. Walter, as well as the rest of us who worked on the study with him, felt the elderly should not be treated with kid gloves. Just like younger people, they had to make a concerted effort in order to realize any significant strength gains.

Our 12 older subjects—ages 60 to 72—were encouraged to train at 80 percent of their one repetition maximum. ("One repetition maximum"—or 1RM—is defined as the most weight a person can lift with one try.) Our troupers worked out for 12 consecutive weeks, 3 days a week, under our watchful eyes. Two muscle groups were the object of our scrutiny: the extensor muscles that extend the knee, known as the "quadriceps" muscles; and the flexors of the knee, known as the "hamstrings."

Besides the intensity of the men's exercise efforts, there was another key difference in our approach: The amount of weight the men lifted was adjusted upward every week. Because the men were getting stronger all the time, we reasoned that the weight had to be raised to make them continue to lift 80 percent of their 1RM.

The results of the study should make us all revise our thinking about strength-building among senior citizens. The strength of the men's quadriceps more than doubled and the hamstrings tripled. The average *daily* increase in the extensors was 3.3 percent and the flexors 6.5 percent.

To what did we attribute this marked increase in muscular strength?

We used two sophisticated techniques to asses the changes in muscle size—and to decide if they were, indeed, growing.

The first is computed tomography, or the CT scan (see figure 2-1), which gives a very precise measurement of total muscle size. Using this method, we observed a remarkable 12 percent collective increase in the men's muscles.

The second is muscle biopsy to permit a size measure on a cellular level. A small piece of muscle tissue was extracted from the subjects before the training, after six weeks, and at the end. Examining the tissue samples under a microscope, we could see that the muscle cell size had increased at each juncture of the study.

Yes, hypertrophy—muscle growth—had occurred. It was visible both microscopically and to the eye. Our conclusion was that *the amount of hypertrophy was as much as we could expect to see in young people doing the same amount of exercise.*[6]

We weren't the only ones who were impressed with the results. The men in the study were amazed at their own improvement. By the end of the training, many of the men were able to lift more weight than the 25-year-old graduate students working in our laboratory. They were so enthusiastic about their progress that many continued their training program after they left us.

More recently, Maria A. Fiatarone, M.D., joined our laboratory. As a specialist in geriatric medicine with an interest in the effects of training on the very old, Maria decided to expand on our earlier study, only her subjects would be the frail, institutionalized elderly. Her goal was to help such people improve the quality of their lives by making them stronger, thereby increasing their functional capacity.

Her initial study was conducted at the Hebrew Rehabilitation Center for the Aged, a chronic-care hospital with more than 700 patients. Her group consisted of ten men and women, ranging in age from 87 to 96 years old. Just as Walter did, Maria had her subjects train at 80 percent of their 1RM. The training period was eight weeks.

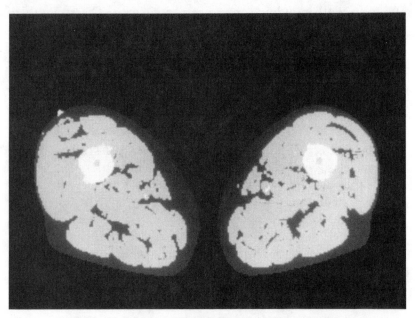

Figure 2-1 **THE CT SCAN: MEASURING MUSCLE SIZE**
Computerized tomography (CT) gives us a picture of what the inside of the leg looks like. It has become one of the most valuable tools we have to evaluate how the individual components of the leg respond to strength training.

These computer-enhanced images show the cross section of the thighs of a man in one of our strength-training studies, as seen by the CT scanner. The darker area is fat; the lighter area is muscle; the white bone is in the middle.

Maria's focus was somewhat different. She was studying the relationship between the participants' muscle strength and how fast they could walk based on the premise that the weaker a person's legs are, the longer it will take him or her to walk 20 feet. Walking time, thus, is a functional way to measure leg strength.

The findings of this study were even more startling. Once again, we analyzed changes in total muscle size using the CT scan. As figure 2-2 indicates, the subjects' muscle strength almost tripled and the size of their thigh muscles grew by more than 10 percent.[7]

Sam Semansky, a 93-year-old study participant, spoke for many of his peers when he observed, "I feel as though I were 50 again. Now, I get up in the middle of the night and I can get around without using my walker or turning on the light. The

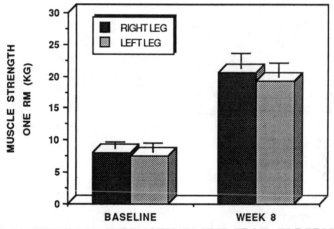

***Figure 2-2* BUILDING STRENGTH IN THE FRAIL ELDERLY**
An 8-week study by Dr. Maria Fiatarone shows that the frail, institutionalized elderly can build strength through exercise, too. Her subjects were 10 men and women—aged 87 to 96—from a midsize chronic-care hospital. Her premise was that the quality of life of the "old old" can be improved by making them stronger, more capable of getting around by themselves, and less prone to falling.

Dr. Fiatarone focused on the relationship between her subjects' leg muscle strength and their ability to walk. She found the stronger they were, the faster they could walk a 20-foot course.

The graph shows how much the subjects' leg strength improved over the course of the study. Leg muscle strength almost tripled and the size of their thigh muscles increased by more than 10 percent. Perhaps as important was the psychological effect. The subjects' confidence in their ability to walk—and walk faster than they had only 8 weeks earlier—soared!

program gave me strength I didn't have before. Every day I feel better, more optimistic. Pills won't do for you what exercise does!"

These studies offer pretty conclusive evidence that *muscle mass and strength can be regained, no matter what your age and no matter what the state of your body's musculature before you start your exercise program.* This is precisely what our Biomarkers Program is designed to do for you.

BIOMARKER 3: YOUR BASAL METABOLIC RATE (BMR)

"Metabolism" refers to the chemical processes in your body that build and destroy tissue and release energy, thereby generating heat. "Basal" means at baseline or at rest. So your basal metabolism is the rate of your body chemistry when your exertion is minimal. It's usually measured just as you wake from sleep, which is about the same as when you're sprawled out on the sofa watching television.

When you eat food, your metabolism goes to work on the food calories, breaking them down and eventually releasing the energy in the form of work or heat. Of course, the more you exert yourself, the more calories your body needs to fuel it. The more you lie around and do nothing, the fewer calories you require. But even at rest, it should be pointed out, your body consumes—or "expends"—some calories.

So what does all this have to do with aging?

Your basal metabolic rate (BMR)—or caloric expenditure at rest— falls with age. Study after study has tried to quantify the extent of this BMR slowdown. Researchers always came up with widely divergent figures until they realized that a person's lean-body mass is the key to compiling a definitive figure.

Here's the reason:

If you have a reduced amount of muscle, as most middle-aged people do, your metabolic demand for oxygen during rest declines, as does your caloric need. Muscle is active tissue requiring nourishment, after all. Fat is more passive tissue that just sits there acting as a storage form of body energy. *We feel that older people's reduced muscle mass is almost wholly responsible for the gradual reduction of their basal metabolic rate.*

Based on our estimates of the average loss of lean-body mass with age, *a person's basal metabolism drops about 2 percent per decade starting at age 20.*

In affluent countries like America where food is abundant, middle-aged people fight the battle of the bulge. It's caused by a little understood, vicious cycle. As people's basal metabolism falls, their food calorie need falls with it. It may surprise you to learn that *an average 70-year-old person needs 500 fewer calories per day to maintain his or her body weight than an average 25-year-old. The average 80-year-old needs 600 fewer calories.* In short, from around age 20 onward, you need to take in about 100 calories per day less each decade to maintain the status quo.

The problem is that many middle-aged people, who require fewer and fewer calories as time marches onward, continue eating as if they were still 20-year-olds. Too many calories coupled with too little exertion, a reduced musculature, and a declining metabolic rate add up to more and more fat. This is why this cycle—unless broken by a program such as ours that increases muscle and restores lost BMR—will only worsen over time.

BIOMARKER 4: YOUR BODY FAT PERCENTAGE

Yes, with advancing age, most of us gain fat *even if our body weight hasn't increased that much.* Our musculature shrinks while fat tissue accumulates, as the two photos (figure 2-3) show so graphically.

Based on our evaluations of hundreds of older people's body makeup, we estimate that the average *sedentary* 65-year-old woman is about 43 percent "adipose tissue," the more scientific term for fat. Contrast that with the average 25-year-old woman's body fat; it hovers around 25 percent. Men, by nature, remain somewhat leaner even as they age. For a man, we see average body fat of 18 percent at age 25, moving up to 38 percent at age 65.

To study body composition, scientists use the ratio between a person's lean-body mass and fat tissue. It's called "body fat ratio." The ratio of the good lean-body mass to unwanted fat decreases as we age.

Don't Let That Proverbial Preoccupation with Poundage Lead You Astray

It's been our observation that most people who undertake an exercise program do so to lose weight, even though for people 45 and older there are plenty of reasons other than simply slimming down to take up exercise. Unfortunately, too many older people

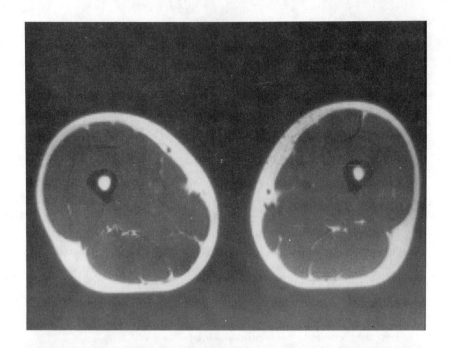

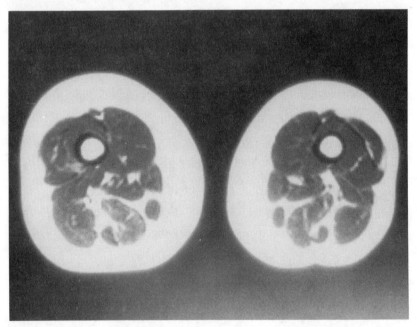

Figure 2-3 **PROPORTION OF LEAN-BODY MASS VERSUS FAT IN YOUNG AND OLDER WOMEN**

These two magnetic resonance images make a dramatic point about the loss of lean-body mass and the accumulation of fat as we age. Both show a cross-sectional view of a woman's thigh. The top photo is the thigh of a 20-year-old athlete. The bottom is the thigh of a sedentary 64-year-old woman. The young woman has a body-mass index (BMI) of 22.6; the older women, a BMI of 30.7.

remain preoccupied with the numbers on their bathroom scale. This is a serious mistake, as we already mentioned. For one thing, muscle tissue weighs more than body fat. That means that a person covered with muscle—a body builder or marathon runner, let's say—could easily weigh more than someone who is largely flab.

Losing weight is the wrong goal. You should forget about your weight and instead concentrate on shedding fat and gaining muscle.

We must warn you: Should you succeed at our goal, you could find that you haven't lost that much weight and that you're out of sync with what those weight tables indicate for a person of your sex and height. Still, you would be much healthier as a consequence because you've altered your body composition in a favorable way. You've decreased your risk for developing sarcopenia.

BMI—Relating Chronic Disease Risk to Weight

So you've decreased your risk for sarcopenia, which fortunately is not a chronic condition if you're willing to follow our Biomarkers Program. However, you may be wondering if there's a way to assess your risk for developing some of the chronic diseases associated with aging. The answer is "yes"—by utilizing something called a "Body-Mass Index" (BMI).

BMI is a way to judge weight in relation to your height. Epidemiological researchers use it to help predict someone's risk for developing a chronic disease.[8] Your BMI is easy to calculate using the nomogram on the following page.

Notice that this nomogram consists of three vertical scales. To calculate your BMI, do three things: (1) Put a dot indicating your weight on the left scale. (2) Put a dot indicating your height on the right scale. (3) Connect the two dots with a straight edge and note the number on the middle scale where your line intersects it. That's your BMI.

For example, suppose you're a man who is 5 feet 2 inches and you weigh 170 pounds. Your BMI is 30.5 and you'd be considered obese. However, if another man who weighs the same amount is 6 feet tall, his BMI is 20 and he'd be very close to an ideal body weight.

An aside: *Weight and height estimates aren't good enough when you're figuring your BMI!* We implore you to be completely honest and use your real weight and height to figure our BMI. We've seen too many prospective subjects for our exercise studies distort their true weight and height when we ask for these figures. It's amazing how consistent people are about saying they weigh less than their actual measured weight and that they're taller than their real height. If you're 5 feet 11 inches tall, you're not 6 feet, nor is 143 pounds the same as 140 pounds. Any small distortion like this becomes magnified when you plug the figures into the BMI nomogram.

BODY-MASS INDEX NOMOGRAM

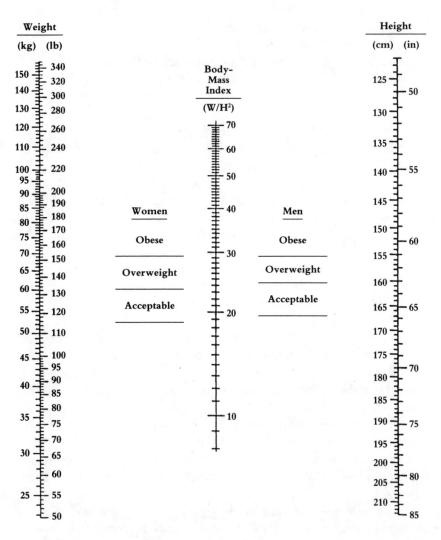

Source: G.A. Bray, *Obesity in America,* NIH Publication, No. 79–359. November 1979, p. 6.

Interpreting the Meaning of Your BMI

Your BMI can be extremely revealing if you know how to interpret it. Your BMI can do two things—help you determine your risk of dying before you reach your full life expectancy, and, at a glance, give you an estimate of how much weight you need to lose.

For years, investigators have been using BMI figures to relate a person's weight to research findings from large-scale epidemiologic studies. We have done this in the chart below, where we relate BMI figures to the reasons for premature death. The chart shows the ideal BMIs for men and women in various age groups. *We define "ideal BMI" as that BMI associated with the lowest risk of chronic disease or mortality.*[9]

Age Range	Ideal Body-Mass Index	
	Male	Female
20–29	21.4	19.5
30–39	21.6	23.4
40–49	22.9	23.2
50–59	25.8	25.2
60–69	26.6	27.3
70–79	27	27.8

If your BMI value is 20 percent or more above these ideal BMI values, consider yourself obese. You have a weight problem serious enough to increase your risk of dying younger than you should according to current mortality rates. If your BMI is 20 percent or more below, you also have an increased mortality risk.

By referring back to the nomogram, you can also get a quick fix on what your weight should be. Place a dot on your height on the right vertical scale, then put a second dot in the middle scale on the BMI that's ideal for your age and sex. Connect both dots with a straight edge and see where it intersects the left-hand weight scale. That's what your weight would be at your ideal BMI.

There's a caveat to this. What BMI does not tell you is your percent body fat. This is the key thing to know since, as you know, it's high body fat, rather than too much weight per se, that can endanger your health. For example, many football players have a BMI well over 30, but *thanks to their low body fat levels, they are not at increased risk of premature death because of obesity*. If you're built like a muscular athlete and suspect that your greater-than-ideal weight is not a problem, you should pay particular attention to the section that follows on waist-hip ratio.

Body Fat Distribution: Another Predictor of Disease Risk

BMI is one indicator of your risk for developing a chronic disease and dying prematurely. There's another predictor that may be even better.

Medical researchers, poring over epidemiological findings, noticed that an inordinate number of people with excess fat stored around their waist were more likely to have impaired blood-sugar tolerance—a Biomarker we'll get to in a moment—and, in the extreme, diabetes. Following that lead, they examined body fat distribution on a large population and came to a compelling conclusion:

Body fat distribution may be as important a factor in disease prediction as the percentage of total body mass that's fat. In other words, it's vital to know where fat is stored on your body.

For example, there's increasing evidence suggesting that people who store much of their body fat *above their hips* have a higher risk for developing heart disease, stroke, and diabetes than people who store fat *below their hips*.[10] Respective examples are the pear-shaped woman and the apple-shaped man, opposite.

Why is it healthier to be shaped like a pear than an apple? A recent study by Dr. Richard E. Ostlund, Jr., of the Washington University School of Medicine offers a possible explanation. Cholesterol levels are closely linked to the spots where people carry fat on their torso. In contrast with potbellied people, people with big hips and trim waists have higher levels of HDL-cholesterol, a protective form you'll learn more about when we discuss that Biomarker later in this chapter.[11]

This issue of where fat is stored around the waistline is an independent risk factor for heart disease, stroke, and diabetes. This means that even if you're not overweight but most of your fat is stored around your waist, you have a greater risk for developing one of these diseases. Moreover, if you're a person with a higher-than-ideal BMI *plus* a large bulge of fat around your midriff, you can assume your risk is magnified. You, of all people, are the perfect candidate for our Biomarkers Program.

To measure your waist-to-hip ratio and determine your chronic disease risk, consult Appendix A.

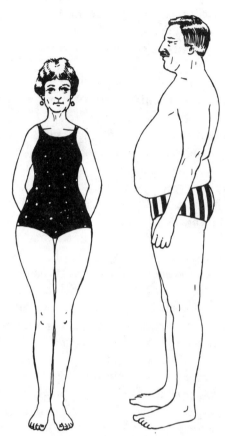

PEAR-SHAPED WOMAN AND APPLE-SHAPED MAN

How Exercise Can Alter Body Fat

Regular exercise plays an important role in maintaining a more youthful-looking body. Everyone knows that. But, you may well ask, to what extent can exercise alter a middle-aged person's body composition if he or she has been sitting on the sidelines for years, losing muscle mass and gaining fat tissue? Can the damage be reversed with an exercise program begun in later life?

It's an important question because a lot of you, reading this book, fall into this category. The answer lies in how long you exercise and what you eat while you're in training.

We're not saying it's easy for older people to lose body fat through exercise alone, although it can be done, as our study showed. To the contrary, *the combination of exercise and moderate caloric restriction is still the best method yet devised to lose weight in a healthful manner.* This is the method we advocate in our Biomarkers Program.

Of all age groups, older people have the most difficult time shedding fat. This is particularly true of aging women. The reason has to do with their basal metabolic rate and their tendency to try to lose weight by diet alone. People trying to lose weight by drastically restricting their caloric intake will likely experience the following adverse consequences:

They'll lose fat, to be sure. But they'll probably lose just as much or more muscle, which, as we've emphasized, is a wholly undesirable outcome. Moreover, a very low calorie diet can also cause deficiencies in a whole host of nutrients, ranging from protein to vitamins and minerals; thus, crash dieters' overall health may suffer. There's another serious impediment to success. Dieters find their basal metabolic rate, which is already lower due to age, being forced even lower by the self-induced starvation process they're undergoing. This is totally counterproductive, for the higher the metabolic rate, the faster fat is burned off. Such people find it a cruel irony that via caloric restriction alone, it becomes even harder to lose weight.

Clearly, *a fat-shedding effort is most productive when it includes exercise.* This is especially true for people whose lifestyles tend to be sedentary—which encompasses a great deal of middle-aged and elderly people. By adding to a weight-loss effort the type of exercise we champion in our Biomarkers Program, you're doing a number of beneficial things:

- You're maintaining muscle mass and losing fat tissue.
- You're raising your metabolic rate and speeding up the fat-burning process.
- You're expending more calories than you take in because exercise consumes extra calories. This means you won't have to restrict your diet radically to the extent that you develop nutritional deficiencies.

BIOMARKER 5: YOUR AEROBIC CAPACITY

Your first reaction may well be, "What is aerobic capacity?" It's a reasonable question, for it's an alien concept to most laypeople.

Aerobic capacity is your body's ability to process oxygen within a given time. That is: (1) to rapidly breathe amounts of air into the lungs for aeration of the blood; (2) to deliver large volumes of blood forcefully via the pumping action of the heart; and (3) to transport effectively oxygen to all parts of your body through the bloodstream. To do these things efficiently, you need healthy lungs, a powerful heart, and a good vascular network. In short, our focus here is on your cardiopulmonary system—your heart, lungs, and circulatory mechanisms. Your muscles figure in here, too, and we'll get to them in a moment.

Scores of cross-sectional and even a few longitudinal studies lead to an inescapable conclusion: *Most people's aerobic capacity declines with age.* Maximum oxygen intake begins to decline at about 20 years of age in men. In women, the decline is often postponed until the early 30s. But in both sexes, *by age 65, aerobic capacity is typically 30 to 40 percent smaller than in young adults.*[12] It's true, however, that the decline is less in older people who exercise regularly.

If you were very active and athletic when you were young and have almost completely ceased exercise in middle age, you'll be disappointed to learn that your earlier sports prowess won't stand you in better stead now. It's a cruel irony that quite the opposite is true, in fact. The loss in your aerobic capacity will be greater than in a person who maintained the same sedentary habits his or her entire life. The reason: The body adapts to the change in the level of activity by making adjustments. The adjustment, in your case, is a very steep drop-off in your once elevated aerobic abilities.

Why does our body's ability to transport oxygen from the atmosphere to working tissues decline as we age? Another way of putting it might be—why does our cardiovascular system become less efficient?

The reasons are numerous, ranging from anatomical changes in the rib cage that adversely affect the chest wall and lungs to deterioration of the heart muscle. Yes, our peak heartbeat—the highest rate a person can achieve—is yet another function that decreases with age, giving rise to the oft cited maximum heart-rate formula: 220 minus your age. (Using this formula, the peak heart rate of a 20-year-old is generally 200 beats per minute. For a 60-year-old, it's closer to 160.)

As the years wear on, our hearts become less responsive to the surge of adrenaline that occurs during exertion; and sedentary people who seldom exercise experience a reduction in heart size. A smaller heart coupled with a reduced maximal heart rate means a corresponding reduction in the maximal cardiac output—which is the amount of blood the heart can pump in a given time period.

Measuring Aerobic Capacity in a Laboratory

Investigators measure aerobic capacity—also called "maximal oxygen intake" or "work capacity"—by asking subjects to exert themselves to the point of exhaustion on a treadmill, stationary bicycle, or some other physical-exertion device. The idea is to place the maximum amount of stress on the system within the confines of safety. While the measurement is taking place, technicians are constantly monitoring subjects' heart rate, blood pressure, and respiratory rate to insure safety.

During the test, subjects breathe into a one-way tube that collects all the expelled air and enables investigators to assess the amount of oxygen their bodies are utilizing during their all-out exercise effort. This oxygen amount is measured in milliliters per kilogram of total body weight per minute, or "ml/kg/min." (Another way of representing maximal oxygen intake is with the symbol "VO_2max.")

A person's aerobic capacity is often viewed as a good index of overall cardiovascular fitness. Indeed, to our way of thinking, it's the best single indicator of fitness and functional capacity there is. It's truly a holistic health measurement.

Bill Foulk, a 56-year-old marathon runner, did better for his age than anyone to date that we've tested in our Tufts Physiology Lab. Bill started running on our treadmill at a speed of 11 m.p.h. It was a comfortable speed for him. Every two minutes we increased the intensity of Bill's effort by steepening the grade of the treadmill.

At the same time, Bill was hooked up to various measuring instruments, including an EKG heart monitor. We were also collecting through a mouthpiece all the air he was breathing. It wasn't long before the dials on the meter that measured Bill's ventilation were spinning wildly. Not long into the test, he was already breathing more than 100 liters of air per minute.

At a 10 percent incline, Bill was still cruising easily, showing no signs of fatigue. At a 12.5 percent grade, his heart rate increased again, as did the volume of air he was breathing. Gradually, as the incline simulated a steeper and steeper hill, it became noticeable Bill was straining. His stride changed. Eventually we could all see he was close to his maximum effort.

"Can you go one more minute?"

Bill nodded his head yes. Now the chorus of voices urging him on increased in intensity. "Bear down, Bill! . . ." "The finish line is in sight! . . ." "Don't give up yet! . . ." "You can do it, Bill!"

At the end of that last minute's worth of supreme effort, the test was stopped and the treadmill was slowed to a 4 m.p.h. walk. At the peak of his performance, Bill was taking in 145 liters of air per minute and his heart rate was approaching 180 beats per minute.

When we removed the mouthpiece, Bill's first words were, "How'd I do? I think I could have gone a little longer."

As it was, Bill's performance was remarkable. His VO_2max. was 72 milliliters of oxygen consumed per kilogram of his body weight per minute. That's a value we only see in highly trained, *young,* endurance athletes.

What a High VO_2max. Indicates

Older athletes like Bill Foulk are proof positive that your aerobic capacity doesn't have to decline with age if you're willing to expend the necessary time and effort to maintain it. We must emphasize the importance of time here because it's an unfortunate fact that *older people have to exercise regularly over a longer period of time to achieve VO_2max. levels equivalent to those of young adults.* Put another way, VO_2max. is a function of the amount of time an aging person is willing to spend working out.

Here's what Bill Foulk's results told us about the condition of his cardiovascular system:

• *Bill's VO_2max. shows how much air his lungs can ventilate.* The more air we can move in and out of our lungs, the better, of course.

In a normal person, as exertion intensifies, the brain stimulates increased huffing and puffing by causing the muscles in the chest wall and diaphragm to expand. This makes it easier for the lungs to move air in and out more rapidly. Unfortunately, in people suffering from asthma, emphysema, or any other chronic obstructive pulmonary disease, the body's ability to respond in this fashion is impaired.

• *Bill's VO_2max. indicates how much oxygen is moving out of his* *lungs into his red blood cells.* "Diffusion" is the term for the movement of oxygen (1) across the lung membrane; (2) through the blood capillary membranes embedded in the lung walls; and (3) into the blood's red blood cells. Diffusion of oxygen into the blood occurs very rapidly. It has to, especially at very high exercise intensities when it's important that the blood leaving the lungs be completely saturated with oxygen. The more oxygen in the blood, the brighter red in color it is.

Unfortunately, the effectiveness of this diffusion process is lessened in people with a lifetime of smoking to their discredit. The accumulated damage can cause their lung membranes and their lungs' blood vessels to thicken. Thickened membranes impede the diffusion of oxygen into the blood. Thus, the blood that leaves this person's lungs is bluish, indicating a low level of oxygen; and there's impaired oxygen delivery to the rest of the body. End result: greatly decreased aerobic capacity (VO_2max.), which places considerable limits on the person's functional capacity.

• *Bill's VO_2max. is related to his heart's ability to pump oxygenated blood to his muscles.* The blood that leaves the lungs flows a short distance into the heart. The heart is little more than a high-powered pump that forces blood to make the long, convoluted journey through the arteries to the rest of the body and back.

The amount of blood the heart can pump is termed "cardiac output." It's a product of how fast your heart can pump (heart rate) and the volume of blood each beat can force out of the heart (stroke volume).

In a healthy person, cardiac output can increase from 5 liters of blood per minute at rest to as much as 20 liters during maximum exertion. However, maximum cardiac output is one function that does wane with age, even in Bill Foulk and other highly trained endurance athletes. So far, research has shown that this *flagging cardiac output is probably an irreversible phenomenon.*

• *Bill's VO_2max. is an indicator of how well blood is reaching his muscles.* Circulating blood, carrying its vital oxygen load, flows from the large arteries into tiny capillaries. Aging and inactivity slow down capillary growth. Fewer capillaries means a lower supply of oxygenated blood can reach muscle tissues.

Whether you're old or young, regular exercise will improve your body's capillary density. This translates into muscles awash with a rich supply of blood during exertion when it's needed most.

• *Bill's VO₂max. indicates how productively his muscles are utilizing oxygen to cause movement.* Your muscles' "oxidative capacity" is critical. This is the ability of the muscle cells to utilize oxygen to convert the energy stored in carbohydrates and fat into a form of energy that supports physical activity.

One of the worst things about inactivity is that it reduces muscle cells' oxidative capacity. This contributes to the extreme muscular fatigue that many aging people experience.

In Older People, Exercise Conditions the Muscles More than the Heart

Regular aerobic exercise—the kind that makes you huff and puff— can bring about a large increase in the muscles' oxidative capacity, especially in older people. Indeed, the latest scientific evidence now suggests that when older men and women exercise aerobically, the muscle cells—not the heart or cardiovascular system—are the things changed by the exertion.[13] Investigators have found that although all of their older subjects increased their aerobic capacity, their maximal cardiac output did not change. Virtually all of the body's adaptations to the rigors of exercise took place in the muscles. *This is the main difference in the way young and older people's bodies are changed for the better in response to aerobic exercise.* (See figure 2-4.) It will be the main difference in how your body responds to our Biomarkers Program compared with that of a younger person.

And that brings us to yet another reason why the strengthening exercises in our Biomarkers Program are so important. They build muscle mass; and when older people have more muscle, it stands to reason that they also have higher aerobic capacity because they have more muscle cells to consume oxygen. When you exercise, you breathe deeply and your heart beats faster for one reason—to pump as much oxygen to your working muscles as possible. The more muscles that are demanding oxygen, the greater your utilization of oxygen and your aerobic capacity. The opposite is true when you are suffering from sarcopenia. When you have a weakened musculature, you have a much lower aerobic capacity because you've got much less muscle tissue demanding oxygen.

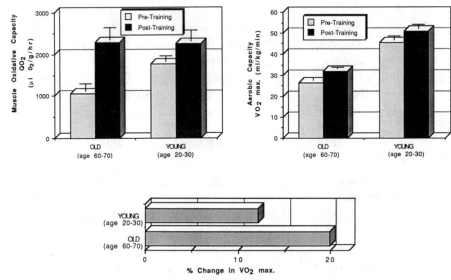

***Figure 2-4* AEROBIC FITNESS IMPROVEMENTS IN
TWO AGE GROUPS**

Depicted in these graphs are the results of a study on the oxidative and aerobic
capacities of two groups—young people (aged 20 to 30) and older people (aged
60 to 70). The study offers convincing evidence that exercise has a greater
impact on the oxidative capacity of middle-aged and older people than it does
on young adults.

In our lab, we took baseline measurements of both groups' aerobic
capacity (see VO_2 max.) and oxidative capacity (see QO_2), which is the
muscles' ability to utilize oxygen to release energy. This was done before the
subjects embarked on a concerted program of exercise training on stationary
bikes 3 days a week, 45 minutes per session, for 12 weeks.

As you can see in the graph at the top left, at the start of the study, the
oxidative capacity of the older subjects was about half that of the younger men
and women. After training, though, the older group's QO_2 measure shot up
even beyond that of the young adults. Indeed, the young group showed only
minimal oxidative capacity improvement.

Contrast these results with the changes in aerobic capacity, which is a
broader, overall measurement of cardiovascular fitness. As the top right graph
shows, the improvement in both groups' aerobic capacity was about the same,
at least in absolute terms. However, when these aerobic capacity improvements
are expressed in percentage terms, the picture changes. The chart on the
bottom makes it clear that the relative improvement was substantially greater
in the older subjects for an obvious reason—their fitness levels were much
lower at the start of the study. They had more to gain from any increase in
aerobic capacity.

Source: Adapted from Meredith et al., "Peripheral Effects of Endurance Training in
Young and Old Subjects," *Journal of Applied Physiology* 66 (1989): 2844–49.

BIOMARKER 6: YOUR BODY'S BLOOD-SUGAR TOLERANCE

The ability of our bodies to control blood sugar (or glucose) is called "glucose tolerance." Unfortunately, with advancing age, our bodies gradually lose the ability to take up and productively use this sugar from our bloodstreams—although most of us don't know it because impaired glucose tolerance, like high blood pressure, has no symptoms. As our blood sugar levels rise with age, our chances of developing "maturity-onset diabetes"—also known as "Type II diabetes"—increase.

By age 70, some 20 percent of men and 30 percent of women have an abnormal glucose tolerance curve, which increases the risk of developing diabetes. Figure 2-5 shows how the incidence of diabetes increases with age in a representative sample of Americans.

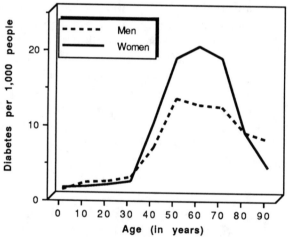

Figure 2-5 **INCIDENCE OF DIABETES BY AGE**
As we grow older, most of us are slowing down, exercising less, and becoming more and more inactive with greater accumulations of body fat. At the same time, the tissues in our bodies are losing their sensitivity to insulin and the cells of our pancreas must produce more and more insulin to have the same effect. Since it's the hormone insulin that helps us metabolize the glucose in our blood, this means our bodies have an increasingly hard time utilizing glucose productively. End result: Our blood-sugar levels rise and we become prone to adult-onset Type II diabetes, also termed "non–insulin dependent diabetes mellitus," or NIDDM.

Source: R. J. Shephard, "Gross Changes of Form and Function," *Physical Activity and Aging,* 2nd ed. (Rockville, Md.: Aspen Publishers, 1987): 129.

Fortunately, researchers are discovering that the reason for this age-related decline usually has more to do with mature adults' higher body fat content and lower muscle mass than with their pancreas's diminished ability to secrete insulin. This offers hope since the amount of fat and muscle on your frame is something you can control through programs such as ours.

By way of background, here's how the blood-sugar drama in your body plays itself out every day:

Dietary sources of glucose are starches—bread, pasta, potato, and so on—and all those items that taste sweet. The process of digestion breaks down carbohydrate starches into individual sugar molecules that enter your bloodstream as glucose.

No matter how much starch we eat or don't eat, out bodies try to maintain blood glucose at relatively constant levels. For example, if our blood glucose level drops too low, our liver steps into the breach and manufactures glucose. The body seeks this glucose constancy for survival reasons. Glucose, after all, is the brain's only fuel, and the brain is the control center of the body.

The average blood-sugar level of a young person we'll call Joe is about 85 milligrams of glucose per 100 milliliters of blood when he wakes up in the morning after a 10-hour overnight fast. After Joe consumes sugar in the form of breakfast cereal, his glucose levels rise, causing his pancreas to swing into action. The pancreas —more specifically the beta cells of the pancreas—secrete the hormone insulin, which immediately begins to stimulate his body's muscle cells to take in and utilize the glucose circulating in his blood. Without insulin doing its job regulating glucose levels, sugar would rise in the blood in an uncontrolled way while body cells would starve.

Muscle tissue is programmed to respond to insulin. It's *sensitive* to insulin. In a normal person like Joe, X amount of insulin stimulates his muscles to take up X amount of glucose.

However, this quantitative relationship is altered for the worse when people age. As they develop too much body fat and too little muscle—sarcopenia—their muscle tissue becomes less and less sensitive to insulin. As a consequence, in many older people, it takes more and more insulin to have the desired effect. This would be fine except for one problem. The production cells in the pancreas, called "beta cells" (b cells), unlike most other types of cells in the body, can eventually burn out from overuse, leaving such people with an impaired ability—or no ability—to manufacture insulin anymore. This condition is called "adult-

onset diabetes" (or Type II) and requires insulin therapy. Fortunately, if Type II diabetes is detected when it's still mild, often
diet and exercise are enough to solve the problem without resort-
ing to drugs or insulin injections.

What's Your Diabetes Risk?

We think *creeping blood-sugar intolerance* is one of the most dev-
astating of the so-called age-related changes. It's certainly one of
the most dangerous because it can turn into diabetes, and uncontrolled diabetes is a killer. It contributes to high blood cholesterol
levels, hypertension, and heart disease.

Where your body fat is stored offers an indication of the extent
of your risk for developing maturity-onset diabetes. It's now
clear, based on laboratory observations and numerous other studies, that upper-body fat—the typical "beer belly" seen on men
and the roll of fat some women develop *above the waistline*—puts
you at greater risk. In Appendix A we explain how to assess your
diabetes risk via your waist-to-hip ratio.

The Impact of Diet and Exercise on Your Risk for Developing Diabetes

While our glucose metabolism seems to change for the worse
as we age, the reason has more to do with other age-related conditions. In other words, aging is an indirect rather than a direct
cause of creeping glucose intolerance—at least, that's the opinion
of Dr. Gerald Reaven, a University of California gerontologist
who has studied the problem in depth. Our research, too, tends
to support this idea.[14]

*The more direct causes of insulin insensitivity are two other factors
associated with aging: increased body fat and inactivity. The two usually
go hand in hand.* A third key cause is a diet rich in fat.

Studies have shown that a diet high in fat, especially in mature
people, is bad, for it tends to lower the body's insulin sensitivity.
In contrast, a diet high in fibrous carbohydrates has the opposite
effect, according to studies by Dr. James W. Anderson of the
University of Kentucky. His research showed that a diet low in
fat and high in fiber can increase the cells' sensitivity to insulin
within only two weeks.[15]

The effect is even greater if a proper diet is combined with
regular workouts. In tandem, a high-fiber diet and exercise—even
in the elderly—can often transform what was previously an insufficient amount of insulin from the pancreas into an adequate

amount. Such a regimen has the added advantage of helping you lower your body fat, which will also foster insulin sensitivity.

Building muscle through weight-lifting exercise is especially critical for lowering your diabetes risk. Here's why:

Muscle, as you've just learned, is the primary site of glucose disposal. Glucose taken up by muscle has three possible fates: It's burned up immediately to release energy; stored as a reserve fuel called "glycogen"; or circulated back to the liver in molecules that can be converted into fat.

If we're physically active, most of the carbohydrates we eat are used right away for energy, which keeps the engine of the body chugging along at a steady clip. The rest is stored as glycogen and becomes a reserve energy source to tap during heavy exertion. But here's the rub: Glycogen stores are constrained by the size of the muscles. This is the primary reason why older people tend to have very low amounts of glycogen in their bodies. You could think of muscles as storage tanks and glycogen as gas. The only way to store more glycogen-gas is to increase the size your muscle-fuel tanks. This is why we are encouraging you, our aging readers, to adopt a strength-building exercise regimen in addition to aerobics. The greater your muscle mass, the more glycogen you'll have in your muscles, thus the more energy you'll have in reserve for periods of exertion.

A third fate can befall the dietary carbohydrate in sedentary people with low muscle mass. When their liver and muscles extract as much glucose as they need, what's left over can turn into fat via a process called "lipogenesis." However, do not infer from this that carbohydrates can make your body fat. Dietary fat is far more of a villain in this regard. As we mentioned, studies by Dr. James Anderson at the University of Kentucky indicate that simply by altering your diet to decrease the percentage of calories that come from fat, while increasing those from carbohydrates, people can realize an improvement in their glucose tolerance. According to Anderson's studies, it's the fiber content of carbohydrates that causes the greatest improvement.

To sum up: Maturing people like you who want to keep their body fat/muscle mass ratio in check and avoid diabetes, should *eat much less dietary fat and more fibrous carbohydrate, such as raw vegetables and whole grains.* All the while, you should be doing strength-building exercises such as those in our Biomarkers Program to increase the capacity of your muscle-fuel tanks. After several weeks following our Biomarkers Program, you'll find you have a healthier glucose metabolism, less body fat, greater muscular strength, and a bigger reserve store for glycogen energy.

Strength-building exercise is a key to regulating your glucose metabolism. Even when exercise does not lower body fat, it's still been shown to increase the muscles' insulin sensitivity. A working muscle, after all, is an entity that requires the energy delivered by glucose to keep going. It has to be able to use glucose if it's going to continue to work at all. Of course, when exercise also results in body fat loss, studies show that the muscles' sensitivity to insulin increases much, much more.

BIOMARKER 7: YOUR CHOLESTEROL/HDL RATIO

In recent years, cholesterol carried the day in the health columns. Without a doubt, it had all other medical topics beat in terms of sheer verbiage in the popular press and on radio and TV talk shows. At the risk of boring you with familiar material, we feel a little background is in order for those of you who have managed to turn a deaf ear to the flap until now—or are still trying to figure out how this cholesterol brouhaha applies to you.

Cholesterol is a fatty substance that's a necessary component of the body. It plays an essential role in the construction of cell membranes and the production of certain sex hormones. It circulates in the bloodstream in association with protein, in combined entities known as "lipoproteins." Under certain circumstances, cholesterol can collect and form deposits in tissues. When this happens in blood vessels, it's called "atherosclerosis," a condition that contributes to the development of heart disease and other circulatory disorders.

Although some high-cholesterol foods, like egg yolk, put cholesterol directly in our system, most cholesterol in our bodies is manufactured by the liver. Indeed, cholesterol is not a required substance in our diet because our body is capable of producing all it needs. But that process is greatly influenced by the fat in our diet. This is something most people fail to realize. That's why, when you're shopping and select a product because the label trumpets "No Cholesterol," you're not getting all the information you need to make a wise health decision. Unless that product is also low in saturated fat, eating it could boost your blood cholesterol level rather than reduce it.

Blood cholesterol's involvement in heart disease is what all the fuss is about. Heart disease is America's number-one killer, claiming one life every 32 seconds and placing a $50 billion annual burden on the nation's economy. In early 1990, a panel representing a broad coalition of American health organizations and government agencies issued a strongly worded report urging all Americans to reduce the fat content of their diets from an average of 37 percent to less than 30 percent of daily calorie intake, predicting that if Americans follow the directive, there could eventually be a 20 percent reduction in heart disease.

We're happy to report that the massive public cholesterol education campaign is working. Actually, it's been working for the last two decades. Beginning in the early 1970s, the average total cholesterol count of the American population has been dropping slowly but steadily. Interestingly, this decline parallels the rise in the fitness boom in this country, lending further credence to the notion that exercise has a beneficial impact on cholesterol levels, a relationship we'll discuss in a moment.

While our collective cholesterol count in this country may be dropping, this is no cause for huge celebration since a person's total cholesterol level isn't the real crux of the issue. It's more complicated than that. It involves the components that constitute the total cholesterol count and the proportion of each in your bloodstream.

There are four major classes of lipoproteins in your blood: (1) chylomicrons; (2) very-low-density lipoproteins (VLDLs); (3) low-density lipoproteins (LDLs); and (4) high-density lipoproteins (HDLs). Although researchers still have much to learn about each, it's already clear that LDL-cholesterol—and VLDL-cholesterol to a much lesser extent—abet heart disease by causing waxy, obstructive, plaque buildup in the coronary arteries of the heart. HDL-cholesterol does the opposite. Among other things, HDL-cholesterol seems to act as a kind of scouring agent, cleansing the arteries of plaque, thus helping to prevent heart disease.

These facts mean only one thing: Your total cholesterol count isn't the issue. A low total cholesterol level—way under 240—offers no guaranteed protection against heart disease. To guard against heart disease, you must raise the HDL-cholesterol and lower the LDL-cholesterol in your blood.

When you have a cholesterol count done, you must always insist on knowing your total cholesterol/HDL ratio. Arithmetically, it looks like this:

$$\frac{\text{Total Cholesterol}}{\text{HDL-Cholesterol}} = \text{Your Cholesterol/HDL Ratio}$$

Example: A 61-year-old man may have a low total cholesterol level of 147, but an HDL-cholesterol count of 22. That makes his ratio nearly 7 (147/22 = 6.68), which is very high.

The total cholesterol/HDL ratio goal for middle-aged and older men and women should be 4.5 or lower.[16]

Essentially, the lower your ratio the better because it indicates that those deleterious LDLs aren't increasing your risk for a heart attack, stroke, or some other life-threatening circulatory event.

Advancing age doesn't seem to have much effect on HDL-cholesterol levels. They tend to remain constant. But there are still age- and sex-related spikes in the total cholesterol level figures. A number of studies show that the total blood cholesterol level of men tends to increase with advancing age until about age 50, when it peaks and starts to decline. In women, this age-related increase continues until they are slightly older.

Since it's only the total cholesterol levels that tend to increase with age and the HDL component remains constant, the conclusion is obvious: It's the harmful LDL and dubious VLDL forms of cholesterol that are increasing. (You're probably wondering why we have to *infer* that these forms of cholesterol are rising. It's because laboratories seldom measure LDLs and VLDLs directly. Instead technicians measure the other blood cholesterol components and do some calculations to arrive at figures for LDL and VLDL.)

To play it safe, your blood lipoproteins should always stay close to these proportions: LDL, the harmful cholesterol, should never exceed 60 to 70 percent of total cholesterol. The helpful HDLs should constitute approximately 20 to 30 percent. And the VLDL component, which isn't all that significant, should hover around 10 to 15 percent.

Lowering Your Cholesterol/HDL Ratio

This is one Biomarker that exercise alone won't remedy because a confluence of factors play a role. Your genetic makeup, diet, exercise, and obesity can all contribute to a dangerous imbalance in your blood's lipoproteins. So can taking birth control pills and smoking.

People who eliminate cholesterol and saturated fat from their diet and let it go at that are doing a good thing, to be sure, but they could be helping themselves even more by simultaneously increasing their activity level. The reason may surprise you:

A diet change can lower only the harmful LDL-cholesterol. It will not raise the amount of helpful HDL-cholesterol in your system.

What will?

According to the latest studies, *the factors responsible for raising HDL-cholesterol are exercise and lowering body fat, as well as quitting smoking and going off birth-control pills.* Studies have also shown that even a small amount of alcohol—about 6 to 8 ounces over a week's time—increases HDL levels in the blood. However, we don't want to emphasize the latter because alcohol is high in non-nutritive calories. Drinking can contribute to weight gain; and people who are overweight are generally people with higher LDL levels.

A number of studies show a causal relationship between aerobic exercise and HDL levels. While the evidence is dramatic in the case of marathon runners who consistently boast extremely high HDL-cholesterol levels, even modest exercisers can achieve the same effect, albeit to a lesser degree.

In one study, a group of male heart attack survivors (average age 53) undertook a walking program for 13 weeks. The program was not strenuous. All they had to do was walk 1.7 miles, three times a week, for a total of 5.1 miles per week. At the end of the study period, their total cholesterol level had actually increased slightly. But not to worry because, proportionately, their HDL levels had risen much more than their LDL levels.[17]

As the extensive medical literature about cholesterol will attest, active people with low levels of body fat, regardless of sex or age, usually have a healthy blood cholesterol balance. Yes, it's important to watch your diet and lower the amount of calories that come from animal-derived foods rich in saturated fat, as we recommend in Part 4 of this book. But it's also important to exercise and alter body composition in the ways we discussed earlier in this chapter. Our Biomarkers Program will enable you to do this.

A final note: Older people should be aware that certain medical conditions and medications can push up LDL-cholesterol levels. They are thyroid disease, certain kidney diseases, diabetes mellitus, and obstructive liver disease. The problematic medications are diuretics and anabolic steroids. A doctor puzzled by an elevated LDL level that won't budge should consider such a causal link and order more specific laboratory tests in order to get a full picture of the problem. In many cases the solution is straightforward: Instead of concentrating on the high LDL level, treat the medical disorder or change medications.

BIOMARKER 8: YOUR BLOOD PRESSURE

In the early 1970s, the U.S. Health and Human Services Administration embarked on an all-out campaign to raise the public's awareness of high blood pressure—the so-called silent killer. Today, most people know that "hypertension" means abnormally high blood pressure. But the average person still knows far too little about its symptoms or effects.

First, the symptoms. Usually there aren't any. But this certainly does not mean that hypertension isn't dangerous. To the contrary, it's a killer just as the government says—something your doctor already told you if you're one of the 60 million Americans who suffer from hypertension. Hypertension is heavily implicated in strokes and heart attacks, among other serious maladies.

The causes of elevated blood pressure include a hereditary disposition for the problem; obesity; too high a consumption of fat, salt, and alcohol; smoking; and too little exercise.

Fortunately, most of these are things we can control with willpower and self-discipline. Unfortunately, race is also an important indicator of hypertension risk—and nobody can modify that. Blacks are far more prone to age-related high blood pressure than whites.

A large number of communities and populations around the world show no increase in blood pressure with age. The United States is not one of them. Aging white Americans can point to excess poundage and deleterious health habits as the triggers for their blood pressure problems; and black Americans can add genetic predisposition to the list.

A blood pressure reading has two numerical components: The systolic figure, which always comes first, is the pressure that blood exerts on the walls of your arteries when your heart contracts. Diastolic refers to the pressure exerted between heartbeats, when your heart is taking a tiny rest. If the diastolic pressure is elevated, clearly your arteries never get a proper rest. They remain unduly stressed all the time. Over time this constant pressure will cause the walls, mostly of your smaller arteries, to thicken and stiffen.

Normal blood pressure is less than 140 systolic and less than 85 diastolic.

One cautionary note about blood pressure readings. You need a minimum of three readings, taken on several occasions, to get a definitive picture of your blood pressure. One high reading could indicate merely that you've had an especially hectic day. Moreover, each reading must be done properly. Before the reading, you should sit quietly for five minutes. The technician or nurse should make sure your upper arm is bare and extended before positioning the cuff. When the reading is taken, he or she should be supporting your cuffed arm at heart level.

While it's true that black people have a strong tendency for hypertension, race is by no means destiny in this regard. To find out how much of a role nongenetic factors play in the development of high blood pressure, black American men have been compared with the Kung bushmen in northern Botswana, Africa.[18] Both the systolic and diastolic pressures of black Africans are slightly higher in the adolescent years, but from then on the two populations diverge markedly. Both populations had age-related increases in the systolic pressure from age 20 onward, but the levels for African men didn't even break through the normal range until after the seventh decade of life. As for Africans' diastolic pressure, there was almost no change.

However, American blacks showed consistently higher readings in both systolic and diastolic pressures after age 20. By the time black Americans hit 50, they were experiencing signs of hypertension. Since the two populations shared certain racial characteristics, the divergence is probably attributable to lifestyle and environmental factors, such things as diet, exercise, smoking, and drinking habits.

Lowering Blood Pressure Through Diet and Exercise

Many people believe that a low-salt diet is the first line of defense against hypertension. However, only 10 percent of the general population are truly salt responsive in that their blood pressure responds to their intake of salt. Fewer people realize that regular exercise can be extremely helpful in this regard, too. A number of studies suggest vigorous workouts may be an effective way to both prevent and treat high blood pressure. For example, scientists at Dallas's Cooper Clinic Institute for Aerobics Research found that people who maintain their fitness have a 34 percent lower risk of developing hypertension.[19] Other investigators have found that even those who already have high blood pressure can lower it for at least one hour following a single bout of exercise.[20] Consider these facts just one more reason for undertaking our Biomarkers Program.

BIOMARKER 9: YOUR BONE DENSITY

There is an age-related decline in the mineral content of bones that leaves an older person with a weaker, less dense, more brittle skeleton. The cause of this is still unclear. But poor dietary habits, hormonal changes in women after menopause, deficient calcium absorption, and a sedentary lifestyle certainly play key roles. Caucasians and Orientals are more prone to bone loss than black people.

When this bone mineral loss reaches the point where there's a substantial increase in fracture risk, it's called "osteoporosis." However, it's important to understand that osteoporosis is not a necessary or normal component of aging. And contrary to popular belief, it's a disease condition that targets men as well as women. While the condition is more common in postmenopausal women, mature men are victims, too.

Maybe you've wondered why the elderly seem to fall a lot and break bones. Older people fall more than the rest of us for many reasons, including failing eyesight, more foot shuffling and less leg lift or spring to their step, and a greater difficulty in restoring balance once they do stumble. When they do take a tumble, it doesn't take much of an impact for them to sustain fractures. Brittle bones, after all, require minimal force to snap.

For young adults, bone fractures are painful and annoying, but full recovery is largely assured. In old people, bone fractures can be life-threatening. Broken bones are the leading cause of accidental death in the frail elderly. Fractures of the hipbone are particularly lethal. In very old people, one in three women and one in six men suffer hip fractures. Fifteen percent of these hip-fracture victims will die as a result, and another 50 percent will require expensive long-term care.

Some background: The normal adult skeleton is in a dynamic state—that is, it's constantly being maintained by a recurring cycle of remodeling ("resorption") and formation ("calcification"). The degree to which these opposing processes are in balance is influenced by the availability of calcium in the bloodstream and the ability of your bone tissue to utilize it.

Research has shown that, on average, an individual experiences an approximate one percent loss of bone mass per year. Given the current state of research on this subject, it's still difficult to pinpoint the most important reason for this loss, although one recent study indicates that consuming the Recommended Dietary Allowance (RDA) for calcium can slow down bone mineral loss. We do know that most women do not exercise on a regular basis or consume the ideal of 800 milligrams per day of calcium (the average is closer to 500 milligrams per day). Fortunately, both of these deficiencies are relatively easy to correct.

There are two types of bones in our body: (1) the compact, cortical bone, and (2) the spongy, trabecular bone. Cortical bone looks solid, and its spaces can be seen only with a microscope. The bone of your forearm and shin are mostly cortical. The trabecular bone—found predominantly in the spine, hip, and thigh—consists of a three-dimensional lattice of branching, bony spicules. To the naked eye it looks like a sponge with the spaces filled with bone marrow.

More trabecular bone is lost with aging. This explains why the sites of osteoporosis-related fractures are usually the vertebrae of the spine, the femur (thighbone), and the forearm.

Bone loss in men and women varies by degree as well as location. As you probably know, the rate of bone loss is slower in men than in women. For every gram of bone mineral that a woman loses, a man loses only .66 grams.

As for location, over the course of a lifetime the average woman experiences an overall decrease in the neck of the femur —the top section that joins with the hip—of 55 percent. The figure is minus 42 percent for the lower or lumbar spine, the part of the spine at the base or small of the back. For men, the mineral content loss in the femur is only two-thirds that of women. The loss in the lumbar spine is even less—one-quarter that of women. This helps explain why the female/male ratio for those lethal hip fractures is 2:1 and 8:1 for vertebral fractures of the spine.

For women, menopause—or a hysterectomy or resection of the ovaries, tantamount to menopause—is the pivotal event. Before menopause, women's rate of average loss in cortical bone remains low—about 0.3 percent a year. However, within the first five years after menopause, when the body's estrogen production drops precipitously, the rate accelerates to 2.5 to 3 percent per year. In some 15 percent of postmenopausal women, there is also an accelerated loss of trabecular bone to about 6 percent a year. These are the women at greatest risk for osteoporosis.

How Diet and Exercise Can Influence Bone Density

In recent years, osteoporosis has been given big play in popular health magazines, particularly in publications with an overwhelmingly female audience. By now, the message that we lose bone density with age has been trumpeted far and wide. While people recognize the problem, few understand what to do about it. Increasing calcium intake alone, mostly via supplements, is rarely enough.

In truth, the scientific jury is still out on this calcium-consumption approach to the problem of preventing bone loss. While some scientists recommend that all postmenopausal women consume up to 1,500 milligrams of calcium per day—almost twice the National Academy's Recommended Dietary Allowance (RDA) of 800 milligrams—a growing body of research suggests only those women who are consuming low amounts of dietary calcium should up their intake.

A bone-mineral study at our research center came to a similar conclusion.[21] Dr. Bess Dawson-Hughes examined the rate of bone loss in women with contrasting dietary calcium intakes. She found it was only in women eating less than 500 milligrams of calcium per day that an accelerated rate of bone loss occurred. If those women then increased their calcium intake to an amount above the RDA level of 800 milligrams per day, this decline in bone mass was retarded.

There's another remedy for osteoporosis that's received less attention but that we think may be the more productive route, no matter what the calcium intake: the exercise prescription.

Those bed rest studies we mentioned in this book's Introduction provide strong evidence in favor of weight-bearing exercise as a preventive measure. Research shows that two weeks of complete bed rest can cause as much calcium loss from bones as one whole year's worth of aging. Or, to put it another way, the rate of bone mineral loss increases 50-fold during prolonged bed rest. However, when bed-rested patients are made to stand for some time each day, even if they don't walk, their accelerated bone calcium loss stops. Why? Apparently the stress that gravity exerts on bones helps maintain mineral content. This effect holds even if these same patients are allowed to stay in bed for the remainder of the 24-hour period.[22]

Today it's accepted within the medical community that stress placed on a bone repeatedly causes it to get stronger, rather than weakening it, as you might think. It's why the arm that people use to swing a tennis racket has stronger bones than their other arm. In the same vein, a number of studies show that *weight-bearing exercise (such as walking, running, and cycling), continued over an 8-to-24-month time span, can effectively reduce the rate of bone loss.* This is true even in the population at the most risk—postmenopausal women.

At Tufts we targeted this group in a second bone mineral study.[23] The women fell into two groups: well-trained runners and nonexercisers. Leanness is a risk factor for osteoporosis, and our runners were, on average, about 20 pounds lighter than the sedentary women. Nonetheless, their bones were still stronger. Oddly enough, the runners' forearms, a non-weight-bearing bone, were also denser. This makes us wonder if stress placed repeatedly on bone anywhere in the body causes the whole skeletal system to gain mineral content. In short, does exercise exert a whole-body effect on bone? We think the answer is yes.

This whole-body effect from our data suggests that *exercise may help foster the body's calcium absorption.* In this study we found the following: (1) The blood of the exercising women had higher concentrations of the active form of vitamin D—or 1,25-dihydroxyvitamin D. This activated vitamin D aids in calcium absorption. (2) The trained women ate more carbohydrate, which is known to increase the body's calcium absorption. (3) The trained women had more growth hormone, which might increase calcium absorption and make bones stronger.

Another Tufts study examined the bone health of older women before and after a one-year-long exercise training program.[24] The women were divided into four groups:

Group 1 walked at a brisk pace for 45 minutes, 4 days a week. They also drank a dietary calcium supplement, bringing their total daily calcium intake to 1,200 milligrams.

Group 2 walked with the Group 1 women but imbibed a placebo drink so that their daily calcium intake was only 600 milligrams.

Group 3 did no exercise, but they consumed the same high-calcium drink as Group 1.

Group 4 remained sedentary and drank the placebo.

It was our hypothesis that there would be a significant inter-action between exercise and the higher calcium intake. We were wrong. The calcium supplement had virtually no effect. All the active women, even those with lower calcium intake, increased their bone mineral content. The sedentary women were all vic-tims of demineralization.

Dr. Everett Smith at the University of Wisconsin performed a similar study, but over a three-year period. His results were the same as ours at Tufts, except the bone-mineral gain and loss was magnified by the longer time period.[25]

We think the evidence is strong that a brisk daily walk is one critical ingredient for preventing the development of osteoporosis.

BIOMARKER 10: YOUR BODY'S ABILITY TO REGULATE ITS INTERNAL TEMPERATURE

Our bodies come with a built-in thermostat. When we're ex-posed to heat or cold—or exercise in either of these extreme con-ditions—our bodies adjust the internal temperature so that it stays within a degree of 98.6 Fahrenheit or 37.5 centigrade.

Let's take the case of exercising in hot weather. Our principal cooling-off mechanism is sweat. The evaporation of sweat off the surface of our skin is our body's way of maintaining our internal body temperature at safe, sustainable levels. It requires adequate blood flow and water utilization.

Heat-related injuries and dehydration are common among the elderly because *the body's vital thermoregulatory ability diminishes with age.* Both hot and cold weather pose a danger to the aged. Take wintry weather. In young people who are exposed to cold, the body's core temperature stays the same or increases slightly. Under the same conditions, older people sometimes find their body temperature actually falls. The reasons are age-related, due to a lower metabolic rate, which we discussed earlier, and a *less-ened ability to shiver.* Shivering is the opposite of sweating. It's the body's way of generating heat (known as "thermogenesis").

There's a complex of factors responsible for older people's thermoregulatory problems. For one thing, *older people have a reduced sensation of thirst.* Whereas a certain amount of exercise would make a younger person crave a drink, it's unlikely the same amount of exercise would send thirst signals to an older person.

One study compared a group of healthy elderly men to healthy young men of the same weight. After 24 hours of no fluid intake, the elderly group still had little sensation of thirst despite being given salty foods, while the younger men were parched.[26]

In general, older people don't drink enough water. Even when they exert themselves in the heat, often they still fail to replenish their body's water stores in sufficient quantities. This is why we have reminders about drinking water strewn throughout our BioAction exercise plans.

But the problem is more complicated than simply inadequate fluid intake. *It takes a warmer internal temperature to make an older person sweat.* Also, older exercisers tend to have a lower heart rate response and a smaller heart stroke volume than younger people, as we pointed out in the aerobic capacity section. This reduced cardiac output impairs the blood flow to the skin, where the heat building up in the body finds its escape route through sweat. This means heat remains trapped inside the body, so that the internal temperature can soar to dangerous, even fatal levels.

Reduced renal (kidney) function and an impaired ability to concentrate urine also have a lot to do with older people's dehydration and thermoregulatory problems. By the time most people are 70 years old, their kidneys filter waste out of the blood only half as fast as they could at age 30.

The purpose of the kidneys, as you know, is to rid the body of waste and control the internal water balance. In a young person, the kidneys do an exquisite job of this. The kidneys know to produce only a small amount of urine when a young person is dehydrated, in order to conserve internal water stores, and to produce a large amount of urine when the body is awash in fluid. The young kidneys do something else that's extremely vital: Since passing waste material in urine is a key function, the kidneys include the same amount of waste in a small amount of urine as a large amount. This is what we mean by "the ability to concentrate urine."

In old people, the kidney's regulatory mechanism is far less efficient and sensitive to changes in the body's fluid intake. During periods of dehydration, for example, their kidneys still produce large amounts of urine, only it's urine carrying little waste material. In short, too much water is constantly exiting an elderly person's body and it doesn't contain a great enough concentration of waste.

We feel that the elderly's lower fitness—and, in turn, low aerobic capacity—is also implicated in the problem. Certainly low fitness and aerobic capacity are responsible for the fact that *older people have a lower rate of sweating.* Our Biomarkers exercise program can help in this regard. As you gain fitness through regular exercise, you'll be helping your body restore its ability to properly regulate its internal temperature.

Exercise Can Help Repair Your Body's Temperature Control Mechanism

Yes, as an older person, you have a reduced sensation of thirst and a lessened ability to sweat and shiver. While exercise probably won't alter your thirst mechanism for the better, it may still reduce your risk for dehydration injuries. Here's why:

Staying in shape, no matter what your age, helps you adjust to the heat. People of any age who are in shape . . .

• have higher total body water content. Regular exercise increases the amount of water in your blood. This simply means that you can afford to lose more water via sweating without becoming dehydrated.

• sweat more when they work out in the heat. An increased rate of sweating is a good thing because it's the body's primary method of cooling itself, of lowering its internal temperature to acceptable, nonlethal levels.

• lose fewer "electrolytes." These are salts (potassium, sodium, and chloride). Fit people have much more diluted sweat, so they can sweat more without lowering their body's stores of these important minerals.

This does not mean you can exercise with no qualms about dehydration. To the contrary, as an older person you must remain ever vigilant in this regard. The rule of thumb is this: The older you are, the more fluids you must force yourself to drink even when you're not thirsty. In truth, even most young people do not drink enough fluids during exercise.

Before we leave the subject of Biomarkers, we'd like to set the record straight about the so-called anemia of old age. This is not a Biomarker. *While it's true that many elderly people are anemic, it's not because of their age.* Data from several sources, including the famous Framingham longitudinal heart study, indicate that the average, self-sufficient, noninstitutionalized senior citizen experiences no change with age in the number of their red blood cells. Because these cells are the blood's oxygen carriers, a low count contributes to low aerobic capacity and functional capacity.

There are any number of reasons why an older person might have a low number of red blood cells. They include nutrient deficiency and blood loss, even from an intestinal polyp or tumor. A physician will need to determine the cause, which will vary from one person to another.

Part II

MEASURING
YOUR BIO-STATUS

___

CHAPTER
3

TWO TESTS TO MEASURE
YOUR BIO-STATUS

✔ **AEROBIC FITNESS SELF-ASSESSMENT**

✔ **MUSCULAR-STRENGTH SELF-ASSESSMENT**

As you learned in Part One, our bodies become more unique as we age. The older a generation of people get, the more biologically diverse they become.

Maybe 75 percent of our readers are still middle-aged, between 50 and 65 years old. However, the fact that they're close in age means very little to us. What we want to know—and what we want you to know—is *your body's age*. Chronology is not destiny. Your body is not necessarily aging on the same timetable as that of your peers. Indeed, different organ systems within your own body may be aging at widely divergent rates. This distinction between your body's *chronological age* and its *biological age* is an important one. It's a concept we return to again and again in this book.

Your body's age cannot be ascertained by counting the grains of sand that have sifted through to the bottom of the hourglass. Nor should you try to guess at it based on the general treatment you've given your anatomical engine. Heredity, after all, is the wild card in the schema. You may have admirable health habits, but the fact that your father died prematurely of heart disease or some other chronic ailment is a key factor that cannot be discounted.

You're curious: "How do I judge how far my body has moved along the aging continuum?"

The measurement of biological age is a calculation that must be done one person at a time, since chronology offers few clues that are worth anything. Your body's age can be estimated through sophisticated medical testing, but that's seldom practical. Few people have the means or inclination. In lieu of this, we've identified two deceptively simple tests that we feel do an excellent job of assessing important components of your biological age. Besides being surprisingly accurate, they can be self-administered. You don't need us in the room with you to find out how fast your body has already journeyed down the road of life.

The first test assesses aerobic capacity. The other monitors strength. In the context of this book, these self-tests serve several purposes. First, we need to know your physical fitness category right now in order to determine which of our two BioAction Exercise Plans you should follow. Second, our tests can be utilized as yardsticks of progress. After you've been following one of our BioAction Plans for a few weeks, you'll find your original test scores extremely useful points of reference. By retaking these tests—particularly the strength test—at regular intervals and then comparing the new results with the old, you'll be able to chart your progress far into the future.

However, before you take either of our self-tests, it's imperative that you do the three things on this checklist:

Before You Take Our Aerobics or Strength Self-Tests . . .
(a checklist)

✓ Complete the Biomarkers Program Screening Questionnaire in the beginning of Part Three. Do not take the self-tests or proceed any further with the program if you answer "yes" to even 1 of our 16 medical questions.

✓ Read the section on stretching in Chapter 4 and familiarize yourself with our stretching regimen. As we indicate there, stretching should always take place near the beginning of a workout, *but it should never be the very first thing you do.* This may be a departure from what you've read in other fitness books. Before you take either of our self-tests, as a form of warm-up, walk around or do some other relaxed form of aerobics for 5 minutes, followed by 5 minutes of stretching.

✓ Read the section on cool-down in Chapter 4. Cool down for 5 minutes after each self-test.

ASSESSING YOUR AEROBIC CAPACITY

Aerobic capacity refers to your body's ability to process oxygen within a given time. We spotlight it here because it's a good overall indicator of cardiovascular stamina, hence your health in general.

Our aerobic capacity test focuses on your heart rate under the duress of exertion. Thus, care must be taken to screen out any conditions that could make your heart respond abnormally. They're outlined in this box, along with a list of the equipment you'll need.

What You Need to Do and Consider before You Take the Aerobic Capacity Tests
(a checklist)

When you take the tests, make sure none of the following are true of you because these are things that can cause your heart to respond abnormally to exertion:

_____ Too little rest for a day or two before the test.

_____ Fatigue because you've recently completed a physically demanding task.

_____ Consumption of a caffeinated beverage within 12 hours of the tests. Caffeine triggers a significant elevation of your heart rate. On the day before the tests, drink decaffeinated coffee or tea or caffeine-free soft drinks.

_____ Smoking within 12 hours of the tests. Smoking elevates your heart rate. It also causes the blood vessels to constrict, thus raising your blood pressure during the test.

_____ Extreme heat and humidity. Exerting yourself under these conditions will cause you to be dehydrated and perform poorly, especially if you're not used to it.

_____ Consumption of alcohol within 24 hours of the test, also because it induces dehydration.

_____ Medication that alters heart rate. Hypertension (high blood pressure) drugs are a perfect example. So-called Beta-blockers keep your heart rate low in response to a variety of stresses, including exercise. If you're on such drugs, there's no way the tests' results will be valid. You should talk to your doctor about any cardiac side effects of medications you're taking.

To make sure your heart is responding normally, take your pulse (see figure 3-1) at rest three or four times within a 10-minute period. If your pulse stays within a 2- or 3-beat-per-minute range, all's well. Proceed with the tests. If it's more erratic than that or your heart rate is declining, wait until it's stabilized before you test yourself.

Also, make sure you drink plenty of water before the tests, whether you're thirsty or not.

Objects you'll need:

_____ Stopwatch or a watch with an easy-to-read second hand. Many digital watches have a stopwatch function that's ideal.

_____ Good walking or running shoes, but not regular leather street shoes. Your shoes should provide support and be comfortable, thus enabling you to walk in them for a sustained period of time.

One-Mile, Walk-for-Time Test

To perform this test, you need to locate a flat, dry, one-mile walking course. If need be, scout the neighborhood in your car, watching the odometer to check what distances constitute a mile. Better yet, use the outdoor track at your local high school or college if you have access to it. You'll find most of these are one-quarter mile around (1,320 feet or 440 yards), but some are one-third of a mile. Check to make sure by calling the school's physical education department and asking what their track's one-lap distance measures.

In order to assess your performance, you need to know how to measure your pulse (instructions appear in figure 3-1), and you need a watch enabling you to time, down to seconds, how long it takes you to walk the one-mile course. It's often easier—and insures accuracy—if you let your partner time you.

Figure 3-1 **HOW TO MEASURE YOUR HEART RATE (PULSE) AFTER EXERCISE**

To assess aerobic capacity, your heart rate response to exercise is the best indicator there is. You can find out how fast your heart is beating at any given time by measuring your pulse rate.

If you've taken a first-aid course, you know how to take your own or somebody else's pulse. If not, here's a short course:

Step 1 Locate points on your body where an artery can be squeezed against a bony surface. These are called "pressure points." The easiest pressure points to use to measure pulse during or immediately after exercise are on your wrist and the side of your temple.

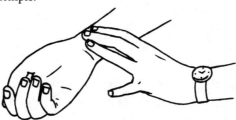

The pulse at your wrist is referred to as the "radial pulse." Hold out one arm—the one without a watch—and use the first three fingers of your other hand as pulse detectors. Put these fingers on the knobby, protruding joint just below your thumb. Slide the fingers inward from there toward the center of your wrist until you come to a pronounced tendon. Gently place the fingers in the hollow just before that tendon. This is right over your radial artery.

We say "gently" because too much pressure cuts off the blood flow. At the other extreme, if you're too gentle, you won't feel a thing. Experiment until you know the amount of pressure to use to pick up the pulsation. Make sure your hand and wrist are loose and relaxed.

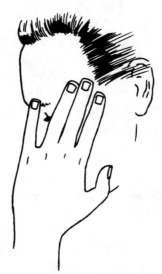

If you prefer to take your "temporal pulse," located on the side of your forehead, that's all right, too. Again, use your first three fingers as sensors and place them on your temple. Let your fingers explore until they find the wide, shallow groove just in front of—and an inch or so above—the ear. The same precautions about gentleness apply.

Step 2 Once you detect a pulse, look at your watch. Start counting heartbeats the moment the sweep-second hand passes any major mark on the dial. End the count after 10 seconds. To get your heart rate per minute, either multiply by 6 or consult this chart, where the calculation is done for you:

12 = 72	16 = 96	20 = 120	24 = 144	28 = 168
13 = 78	17 = 102	21 = 126	25 = 150	29 = 174
14 = 84	18 = 108	22 = 132	26 = 156	30 = 180
15 = 90	19 = 114	23 = 138	27 = 162	31 = 186

This is a walking—not a jogging—test. Walk briskly and try to maintain the pace throughout. The test becomes invalid if, at any point, you start jogging, running, or sprinting. When you cross the finish line, stop and immediately take your pulse (see figure 3-1).

Once you have a beats-per-minute number, consult the appropriate charts for your age and sex in Appendix B.

Here's an example of how a 55-year-old woman might score herself.

The woman takes 15.5 minutes to walk the one-mile course. She checks her pulse at the end and finds her heart rate is up to 155 beats per minute. She locates the chart for her age group and sex (reproduced below) and puts a dot at the 15.5-minute mark along the bottom. Next she traces a straight line up to the diagonal heart-rate grid and puts another dot at 155 beats per minute. To ascertain her maximum aerobic capacity, she simply draws a straight line to the left edge of the chart.

The woman discovers that her aerobic capacity measures 27. In other words, if the woman had come into our lab to have her actual aerobic capacity measured on our treadmill using all of our sophisticated analyzers, the chances are we would have pinpointed her maximal aerobic capacity at 27 milliliters of oxygen per kilogram of body weight per minute (ml/kg/min.).

Granted, that 27 tells you nothing—until you compare it with the norms on the interpretive fitness table (opposite) for your sex. Now you can assess how your aerobic capacity stacks up next to that of your chronological peer group.

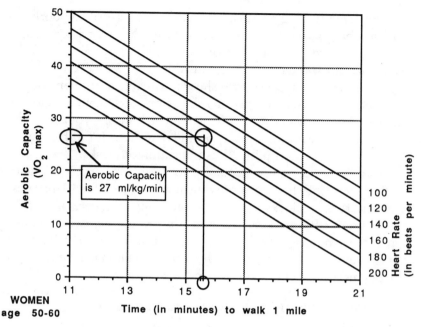

CHART FOR EVALUATING AEROBIC CAPACITY

Walk-for-Time Test
For Men:
Interpretation of Your Aerobic Capacity Score★

Score	Age 40–49	Age 50–59	Age 60–69	Age 70–80
Excellent	50+	46+	41+	36+
Good	41–49	37–45	34–41	31–36
Average	34–40	30–37	27–34	23–30
Below Average	33 and below	29 and below	26 and below	22 and below

Walk-for-Time Test
For Women:
Interpretation of Your Aerobic Capacity Score★

Score	Age 40–49	Age 50–59	Age 60–69	Age 70–80
Excellent	46+	42+	38+	34+
Good	38–45	34–41	30–37	26–33
Average	31–37	27–34	24–30	21–26
Below Average	30 and below	26 and below	23 and below	20 and below

★ The normative ranges on these tables come from our own experiences in the Tufts
physiology lab as well as the wide selection of such values that appear in the medical
literature.

Incidentally, don't be discouraged if you performed poorly on our aerobic capacity test compared with the norms for people your age. Aerobic capacity is one of the key physiological functions that our Biomarkers Program addresses. With exercise training, your aerobic capacity will show noticeable improvement. As your body adapts to regular aerobic workouts, your cardiovascular system becomes more efficient, which means your heart no longer has to beat as frantically at high exertion. It can beat more slowly and still accomplish all its multitudinous tasks.

For example, many people who take our aerobic capacity walking test find that when moving at a brisk pace for 10 minutes, their heart rate clocks 135 beats per minute. Later, after four weeks on our Biomarkers Program, they retest themselves, walking at the same pace for the same amount of time. Voilà! To their amazement, their heart rate reaches only 125 beats per minute this time—and they feel much less winded.

An Alternative Way to Measure Aerobic Capacity: The Step Test

The disadvantage of the Walk-for-Time Test is that it assumes everybody who takes it walks with the same efficiency. That is, the effort expended by a tall, thin person is the same as that of a short, overweight person. Or that a person who has trouble walking because of foot problems or joint pain is on a par with a person who has no handicap.

People with impairments can take the walking test, but they should not compare their results with the norms for their age group. Their test results are valuable for baseline comparison purposes only. When they retest themselves later, they can see how much progress—or lack thereof—they're making.

People with impairments do have another option. In Appendix C we describe a Step Test that is also an excellent way to measure aerobic capacity. It's a more complicated test to take, however. Still, it's a good alternative method for people with foot problems or for anyone who is reading this book in midwinter when icy or snowy groundcover poses an obstacle.

The Strength Test

Like the walking test, this test is also pretty easy to comprehend. Here's all there is to it:

First, you must familiarize yourself with the term *one repetition maximum* (1RM). This is the most weight you can lift with one try. If you try lifting that amount of weight again immediately afterward, you can't because your muscles are too tired. One RM is a measurement of strength we often use in our physiology laboratory.

Unfortunately, 1RM is difficult to measure unless you have a whole panoply of weights to choose from and a professional at your side, guiding your actions. In lieu of that, we're simply going to ask you to lift a weight as many times as you can, and by knowing the number of times you've lifted it—in other words, the number of "repetitions"—you can estimate your 1RM.

Our self-test measures both upper- and lower-body strength.

To test your arm-muscle strength, you'll need either a weight you can grasp or a wrist weight that you strap on.

If you don't want to purchase a weight at your local sporting goods store in order to take this test, a homemade weight is an excellent substitute. You can use any object you find around your house that's both heavy enough to provide a challenge and easy to grasp. For example, a good homemade test weight can be made out of a one-gallon plastic milk or laundry soap/bleach container, the kind with handles. Fill the empty container with water, sand, or lead shot.

Filled with water, a milk container weighs about 8.3 pounds, in our experience, a good starting test weight for women. Men might want to fill the container with sand or lead shot. (You can purchase lead shot at any hardware store.)

If you'd rather use a store-bought weight, we have a few suggestions. Be sure to get a system with a wide enough range of weights because you'll be lifting an increasing amount of weight as you proceed through the Biomarkers Program. Our experience has been that people who follow the strengthening exercise sequence outlined in our BioAction Plans can expect to increase their strength by 100 to 200 percent over a 12-week period. That means a person who takes the Strength Test and finds he can lift only 20 pounds will be able to lift up to 60 pounds 12 weeks from now!

To test your lower-body strength, a commercial ankle weight is probably preferable, although you could make a weight out of a log covered with soft rags.

In preparation for the test, try to select upper- and lower-body weights you'll be able to lift only about 15 times. We've found this self-test to be the most accurate if a person cannot go beyond 15 repetitions. If they can do more than 20 repetitions, the weight is definitely too light and they need to substitute a heavier one.

It goes without saying that *you must know the exact weight of the object you're lifting.* That means weighing your homemade weight on your bathroom scale. Of course, immediately before the test, be sure to warm up and stretch as we described earlier.

Upper-Body Strength Test (Biceps Muscles)

- Sit on a straight-backed chair with your homemade weight on the floor directly in front of you. **Position 1:** Spread your legs and lean forward at the waist. Use your nondominant arm to reach for the weight. (If you're right-handed, use your left arm, and vice versa.) With your arm fully extended and your wrist as straight as possible, grasp the weight. Flex your biceps muscles of your upper arm and use them—*not your back muscles*—to lift the weight. **Position 2:** Bend your elbow and "curl" the weight to shoulder level, roughly to a position in front of your face. **Position 3:** Next, straighten out your arm to a fully extended position, all the while keeping your shoulder steady.

- Keep moving your arm back and forth between Positions 2 and 3 as many times—or "repetitions," in strength-building parlance—as you can. Keep careful count. Do not stop until your muscles absolutely quit on you, when they won't allow you to lift the weight even one more time.
- Consult the "Chart for Predicting 1RM" on page 98. Find the number of repetitions you completed along the bottom of the chart. Trace it straight up on the chart to where it intersects with the solid, black diagonal line. Then, trace it straight left to the numbers on the left border of the chart. Multiply the number

you find there by the amount of weight you could lift. That's your predicted 1RM.

For example, if you could lift 15 pounds 11 times, according to the chart your predicted 1RM is 20.25 pounds (15 lbs X 1.35 = 20.25 lbs).

The amount of weight we'll ask you to lift later on as part of either of our BioAction Plans will be about 80 percent of this 1RM amount. In the above example, for instance, the person's predicted 1RM is 20.25 pounds. Thus, he or she will be lifting about 16.2 pounds.

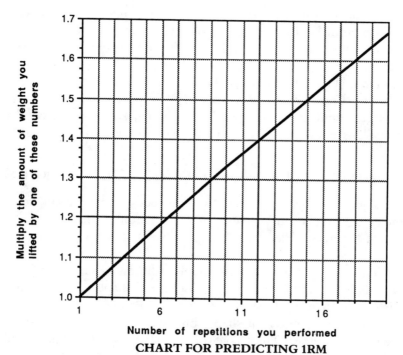

Number of repetitions you performed
CHART FOR PREDICTING 1RM

Lower-Body Strength Test (Quadriceps Muscles)
- Once again, sit on a straight-backed chair. Tie or strap the weight to the ankle of your nondominant leg. (If you're right-handed, your nondominant leg will be the left one, and vice versa.) With the weight nestled in the crook of your ankle, extend your leg so that it's as straight out in front of you as possible. Lower your leg back to its starting position. Rest no more than one second before repeating the same lift, known as a "knee extension."

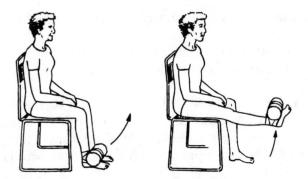

- Lift the weight as many repetitions as possible, keeping count. When your leg muscles are absolutely exhausted, stop.
- Just as you did during the Upper-Body Strength Test, consult the following chart to determine your 1RM.

Don't be surprised if you can't lift as much weight as you thought. Our experience is that very few people actually do any strength training. Therefore, their muscular strength is woefully low. This is often true even in people who are otherwise fairly active.

EVALUATING YOUR OVERALL FITNESS AND STRENGTH

These tests offer a snapshot of your body's strength and aerobic capacity. They also give us the information we need to decide which BioAction Plan you should follow. Circle your test results here:

Those of you who fall into the low-fitness group will undertake the slow-and-easy trip back into good shape by way of our BioAction Plan A, detailed in Chapter 5. People in the medium-fit group aren't in awful shape, but they certainly can't rest on their laurels, either. They have some catching up to do and should follow the exercise program in Chapter 6. Those in the last group are probably able-bodied exercisers already, which would explain why they're in such superb shape. They're destined for the far more sophisticated advice for serious amateur athletes contained in Chapter 9.

Fitness Scorecard

Walk-for-Time Test

| My score: | Excellent | Good | Average | Below Average |

Step Test*

	Phase 1 score:	Phase 2 score:	Phase 3 score:
Excellent	Excellent	Excellent	Excellent
Good	Good	Good	Good
Average	Average	Average	Average
Below Average	Below Average	Below Average	Below Average

Strength Test

	With my arm I can lift . . .			
	Age 40–54	Age 55–69	Age 70+	
Excellent	50+ lbs.	45+ lbs.	36+ lbs.	Excellent
Good	36–50 lbs.	31–45 lbs.	21–35 lbs.	Good
Average	21–35 lbs.	15–30 lbs.	11–20 lbs.	Average
Below Average	0–20 lbs.	0–15 lbs.	0–10 lbs.	Below Average

	With my leg I can lift . . .			
	Age 40–54	Age 55–69	Age 70+	
Excellent	71+ lbs.	61+ lbs.	51+ lbs.	Excellent
Good	51–70 lbs.	41–60 lbs.	31–50 lbs.	Good
Average	30–50 lbs.	21–40 lbs.	16–30 lbs.	Average
Below Average	0–30 lbs.	0–20 lbs.	0–15 lbs.	Below Average

· If the majority of your circles are around "Excellent," you're an exceptional human specimen and, we suspect, a serious athlete. Read Chapter 4 for further background about exercise and then jump to Chapter 9.

· If most of your circles are distributed in the "Excellent" and "Good" categories, you're in pretty good shape. You'll be in even better shape after following our 12-week BioAction Plan B in Chapter 6.

· If your circles are clustered in the "Average" and "Below Average" categories, you're destined for a biological tuneup via Chapter 5's BioAction Plan A for those with low fitness.

· If your results fall at the two extremes—with both "Excellent" and "Below Average" circled several times—we can't make a judgment about your condition, nor can you. There's no distinct pattern. We advise you to retake the tests. Something went awry, which you can fix, we hope, during the second attempt.

---

* The Step Test appears in Appendix C. It's an alternate method for testing the aerobic capacity of people with impairments, replacing the Walk-for-Time Test.

Part III

BIOACTION
EXERCISE PLANS

Part Three of this book is devoted to *ACTION*. The next few chapters focus on exercises that will improve markedly the chances that your body will age at a much slower rate than chronology would normally dictate. By adhering faithfully to one of our BioAction Plans, outlined fully in chapters 5 and 6, you may expect to experience renewed vigor, improved musculature, and more stamina, not to mention new-found flexibility and strength. And that's not all. Many people also report an emotional as well as a physical lift from following our goal-oriented exercise regimen and dietary recommendations. As a consequence, they find their lives becoming more activity-filled and satisfying.

We'll never forget the testament of one such convert, a 58-year-old woman:

"As I moved into my fifties, my body felt more and more like a dead weight. I simply accepted it as a part of growing older. How wonderful to discover this isn't the way I have to feel at my age—to know that, by working at it, I can actually feel better and more ready for life than I did when I was 22!

"Since I've been on the Biomarkers Program, my body feels awake during the day when it should be alert and ready for action. Before, it was as if my body were dulled into a perpetual snooze. I felt lethargic all the time. Now, looking back, I realize how afraid of exercise and immobilized I'd become.

"I'm also proud of the fact that I still have the stick-to-itiveness to stay on a program that isn't always easy or convenient."

Indeed, as this woman points out, the ultimate payoff of our Biomarkers Program is often as much psychological as it is physical. We speak of the strong sense of achievement that people feel when they discover they have the self-discipline to follow through on a productive course of action.

Based on the results of the aerobics and strength self-tests you took in Chapter 3, you now know your baseline fitness. If you're not a regular exerciser, though, there are probably big gaps in your knowledge about how to exercise correctly. Chapter 4 is devoted to filling in those vacuums, giving you all the information you need to begin our program.

There are two BioAction Exercise Plans in this section. The 16-week BioAction Plan A in Chapter 5 is for those of you whose self-test results placed you in the low-fitness group. The 12-week BioAction Plan B in Chapter 6 is designed for moderately fit people. The purpose of both BioAction Plans is to bring both groups of exercisers up to the same fitness level. In other words, when the low-fit people finish their 16-week program, they should be in about the same shape as those moderately fit people who underwent only a 12-week program. *No, you do not graduate from BioAction Plan A to BioAction Plan B.* Instead, graduates of both plans should move on to Chapter 7, which describes the much easier exercise maintenance guidelines to follow for the rest of your life in order to continue the good fitness level you achieved via one of the BioAction Plans.

After firing you up to begin the process of turning back your bio-clock, we're sorry to have to inject a precautionary note at this stage. But we wouldn't be responsible scientists if we didn't. Please complete our medical screening questionnaire in the box below before you take either of the self-tests in Chapter 3 or attempt to start the program.

Biomarkers Program Screening Questionnaire

While our Biomarkers Program is not overly strenuous, there are people reading this book who, because of preexisting medical conditions, should not exert themselves, at least not without securing their physician's express permission. To find out if you're one of them, take the following screening questionnaire developed by Maria Fiatarone, M.D., an HNRCA physician. If you answer "yes" even to 1 of these 16 medical questions, we advise you to have a full-scale medical examination, especially if you haven't had one in a while. After the exam, show a copy of this book to your doctor and ask him or her if it's safe for you to proceed.

	Yes	No
1. Do I get chest pains while at rest and/or during exertion?	____	____
2. If the answer to Question 1 is "yes": Is it true that I haven't had a physician diagnose these pains yet?	____	____
3. Have I ever had a heart attack?	____	____
4. If the answer to Question 3 is "yes": Was my heart attack within the last year?	____	____
5. Do I have high blood pressure?	____	____
6. If you don't know the answer to Question 5, answer this: Was my last blood pressure reading more than 150/100?	____	____
7. Do I have diabetes?	____	____
8. If the answer to Question 7 is "yes": Is my diabetes presently going without treatment?	____	____
9. Am I short of breath after extremely mild exertion and sometimes even at rest or at night in bed?	____	____
10. Do I have any ulcerated wounds or cuts on my feet that don't seem to heal?	____	____
11. Have I lost 10 pounds or more in the past 6 months without trying and to my surprise?	____	____
12. Do I get pain in my buttocks or the back of my legs—in my thighs or calves—when I walk?	____	____
13. While at rest, do I frequently experience fast irregular heartbeats—or, at the other extreme, very slow beats? (While a low heart rate can be a sign of an efficient and well-conditioned heart, a very low rate can also indicate a nearly complete heart block.)	____	____
14. Am I currently being treated for any heart or circulatory condition, such as vascular disease, stroke, angina, hypertension, congestive heart failure, poor circulation to the legs, valvular heart disease, blood clots, or pulmonary disease?	____	____

	Yes	No

15. As an adult, have I ever had a fracture of the hip, spine, or wrist? ____ ____

16. Did I have a fall more than twice in the past year (no matter what the reason)? ____ ____

Even if you checked "no" to all 16 questions, you should be aware that the American College of Sports Medicine urges all people over age 35 who are about to begin a vigorous training effort to have a medical exam. (One caveat: Should your physician discourage you from pursuing strenuous exercise training out of hand, without giving a good reason, we advise you to get a second opinion.)

CHAPTER 4

BASIC EXERCISE CONCEPTS

✓ **STRETCHING EXERCISES**

✓ **WARM-UP AND COOL-DOWN**

✓ **FREQUENCY, DURATION, AND INTENSITY**

✓ **STRENGTH-BUILDING FUNDAMENTALS**

✓ **STRENGTH-BUILDING EXERCISES**

✓ **EXERCISE EQUIPMENT**

✓ **EXERCISE SAFETY GUIDELINES**

––––––

In this chapter, we're going to get you ready to step up to the starting line. But before you step into your workout clothes, there's some background information about exercise that you need to know. In fact, the type of clothes to wear is just one of the topics we'll be covering here. In this chapter you'll also learn the difference between aerobic and isotonic exercise . . . why stretching and an adequate warm-up and cool-down are vital components of any exercise session . . . the concepts of frequency, duration, and intensity . . . muscular-conditioning do's and don'ts, including illustrations of specific strength-building exercise routines . . . the essential and not-so-essential exercise equipment you'll need, both for aerobics and strength-building workouts.

To those of you who've read exercise books before or who

belong to a health club, exercise regularly, and attend clinics, this information may seem like grade-school stuff. However, we urge you to scan through it anyway to make sure you're not missing out on anything.

TWO KEY FORMS OF EXERCISE: AEROBICS AND ISOTONICS

Like most things in life, exercise comes in a variety of shapes and forms. To be sure, all exercise is not the same.

Aerobic exercise—commonly called "aerobics"—is a key component of our Biomarkers Program. It refers to the kind of fast-paced activity that makes you huff and puff. Yes, huffing and puffing is good for you. It places demands on your body's cardio-vascular apparatus and, over time, produces beneficial changes in your respiratory and circulatory systems. Examples of aerobic exercise are brisk walking, jogging, cycling, swimming, re-bounding (running in place on a minitrampoline), aerobic dance, cross-country skiing, skipping rope, and many recreational sports.

Aerobic exercise does not require excessive speed or muscular strength. In contrast, "*an*aerobic" exercise does. As the prefix implies, anaerobic means "without oxygen." Clearly, a living thing can't last long without oxygen, so these are forms of exercise that don't last long either. Sprinting is the best example of anaerobic exercise. You run as fast as you can—so fast you don't breathe—for a matter of seconds, then slow down and breathe as hard as you can to compensate.

Our program does not advocate these short, exhaustive bursts of effort except for serious amateur athletes. No, we most definitely do not want you to model yourself after that showoff hare of Aesop's fables fame. Rather, our program encourages participants to em-ulate the steady, persevering pace of that determined tortoise who, as you'll recall, won the race in the end.

Another form of exercise is isotonic. It's aimed squarely at the muscles and joints. Isotonics encompasses strength-building and flexibility-oriented activities, such as slow stretching and weight lifting. Such exercise requires contraction of a set of muscles, often while you're moving a joint. Some mild recreational sports such as shuffleboard, archery, and horseshoes fall into this cate-gory. But when most people think of isotonics, they envision muscular conditioning and bodybuilding.

Our program encompasses both aerobics and isotonics. They complement each other. And the end result is far greater than what you'd experience from doing only one of the two.

THE 10-MINUTE WARM-UP

Exercise workouts—no matter what the form of exercise— must always begin with five minutes of a low-intensity aerobic warm-up, preferably just walking around. This should be followed by another five minutes of stretching. A gradual warming up of your muscles and ligaments in this way renders them more pliant. Why is this important? Did you ever stretch a rubber band that's really cold? It snaps easily. The same can happen to your body's connective tissue if it's too "cold" because you haven't warmed up properly. Warming up is the way to prevent muscle, tendon, or ligament pulls and tears.

A ten-minute warm-up is an essential component of good exercise protocol whether the workout will involve aerobics or strength building. Its purpose is to gradually increase the body temperature and blood flow to the muscles in preparation for the more vigorous exertion to come:

• For a warm-up, do a slow form of an aerobic exercise for 5 to 10 minutes, followed by 5 minutes of stretching. If you've got time at the end of your session, stretch again for 5 minutes. (Make sure you apply these instructions to the aerobic self-assessment test in Chapter 3.)

STRETCHING THE STIFFNESS OUT OF YOUR BODY

The benefits of stretching are many. Stretching increases flexibility, coordination, and agility and widens the body's freedom of movement. These become desirable goals as we age and many of us grow more sedentary. Too little activity causes muscles to shorten and the joint connective tissue (tendons and ligaments) to weaken. End result: that general stiffness many of us feel. Stretching is a way of combatting these degenerative effects. Over time, stretching lengthens the muscles and creates sturdier tendons and ligaments. By so doing, it helps prevent musculoskeletal injuries during more strenuous forms of exercise, such as aerobics and weight training.

Stretching is also relaxing, mentally as well as physically. We predict it will only take a few stretching sessions to convince you of its value.

Nine basic stretches are shown below. After reading the captions, you'll realize there's nothing fast about this form of movement. Indeed, it's the most languid form of exercise there is. If you're stretching at home, some dreamy music with no discernible beat might help put you in the proper frame of mind. Or if you can't play such music, you could "think" this type of music as you stretch.

Stretch 1. Lying flat on your back, put the soles of your feet together and bring both feet toward your body. Keeping your feet together, spread your knees apart, very slowly, as far as you can. Hold this position for at least 10 seconds and for no more than 20 seconds. Relax and repeat one more time.

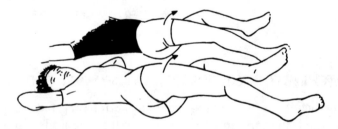

Stretch 2. Staying on your back, bend your left leg and place your left knee sideways on the floor. This will cause the right buttock to rise off the floor. Keeping your shoulders flat on the floor and your right leg straight, roll your buttocks over as far as possible. Hold this position for 10 to 20 seconds, relax, and repeat the stretch. Bend your right leg and place your right knee on the floor. Repeat the stretch for the right side.

Stretch 3. Sit up. Keep your back as straight as possible as you place the soles of your feet together (as you did for Stretch 1). Keep your feet touching the floor and slowly pull your feet toward your body. Hold this position for 15 seconds. Repeat this stretch only once.

Stretch 4. Remain in a sitting position on the floor, your legs extended out in front of you. Keep your back straight as you place the palm of your right hand on the floor behind you, while you simultaneously bend your right knee and cross your right foot over your left leg. Your right foot should be on the floor and your knee bent at about 45 degrees. As you slowly turn your head to the right, use your left arm to push your right leg. Hold this stretch for 20 seconds and relax. Repeat the stretch once as described, then repeat it twice for the left side.

Stretch 5. You're still seated on the floor. Lean back on both arms. Place your left foot behind you with your knee flat on the floor. Your right leg should be bent so that your right foot touches your left knee and your right knee touches the floor. Lean backward and hold the stretch for 20 seconds. Do this three times. Perform three more stretches with the right leg.

Stretch 6. Extend your left leg straight out in front of you. Grab your ankle with your hands, and slowly pull your body forward as far as possible. Try to keep your left knee from bending. It is very important to perform this stretch slowly, with no rocking motion. Hold the stretch for 10 to 15 seconds. Do this twice. Perform two more stretches with the right leg.

Stretch 7. Stand up and face a wall. Place your right palm against the wall for support. Bend your right leg and reach around with your left hand and grab your right ankle. Pull your leg up as far as possible and hold the stretch for 20 seconds. Relax and repeat. Repeat the stretches twice more using the left leg and right arm.

Stretch 8. Find a sturdy desk, chair, or rail bolted to a wall to use as a support. It should be at your mid-thigh level. Keep your back straight as you bring your right leg up and place it on the support. With your left foot flat on the floor, slowly move your body toward the support. The object is to stretch the calf muscles in your left leg as well as the thigh muscles of the right leg. Hold the stretch for 20 seconds and repeat two more times. Repeat three times with the opposite leg.

Stretch 9. Stand about three feet from a wall. Plant your left foot on the floor and take one step forward with your right leg as you cross your arms against the wall and place your head on them. Bend your right knee, keeping your left foot in position behind you on the floor. Move your body down and toward the wall. Hold the stretch for 20 seconds and repeat. Do the same for the right leg.

Stretching is not quite as easy and straightforward as it sounds. Here are some Do's and Don'ts to insure that a stretching session doesn't go awry and actually cause an injury:

Do Stretch

- ever so slowly. Quick, abrupt, jerky movements can cause injury and won't accomplish the task of lengthening your muscles and strengthening joint connective tissue. If you can't reach as far as you'd like, the only thing you might do to earn a few extra inches is rock back and forth very gently.
- by holding each position between 5 and 30 seconds.
- as far as your agility allows. Try to perform as complete a movement as possible, even if it's difficult for you. Over the course of our Biomarkers Program, you'll be surprised at the progress you make toward being able to perform all of the stretches in our illustrations.

Don't Stretch

- by curving your spine. Always keep your back as straight as possible. Back flexion (bending) can trigger an injury in people

prone to back trouble. This can be especially problematic in women with a tendency to osteoporosis.
• if you feel a sharp pain. Stop and give the aching muscle a couple days of rest. If the pain was so sharp that you think you pulled a muscle—or a joint continues to hurt long after your stretching session—consult your doctor or some other trained professional.

THE 5-MINUTE COOL-DOWN

We've just introduced the concept of "warm-up," which always precedes a longer, more intense form of exercise. Its objective is to help your body successfully make the transition from a static, no-activity state to an extremely dynamic state of all-out exertion.

A 5-minute cool-down comes at the end of an exercise session, and it's just as crucial, maybe more so for older adults. Instead of slamming on the brakes after an exercise workout, the cool-down amounts to a sequential down-shifting, from high gear to low and finally to a full stop. You slow your pace in increments until your body is ready for a complete halt.

Here's a description of why your body needs a gradual cooling off; we hope it's memorable enough to convince you *never to forgo it:*

During exercise, your body warms up and starts generating heat. To dissipate the heat, the blood vessels in your skin dilate, accounting for the red face and flushed look many exercisers have. Sweat forms. As it evaporates from the surface of the skin, it cools your overheated body.

This rapid flow of blood to the skin challenges your heart to pump extra hard to keep adequate blood circulating to all the vital organs of the body. Your working muscles actually help your laboring heart accomplish this task.

All is fine as long as you keep exercising and your muscles continue to help your heart shunt the blood freely around your active body. However, you could be in for trouble should you stop moving abruptly. Suddenly your muscles aren't holding up their end of the bargain, and the full burden falls on your heart. Your body temperature is still way above normal, and blood still needs to be diverted to the skin to aid in the cooling-off process. Your overtaxed heart can continue to fulfill its multiple obligations in only one way—by beating even faster.

In this situation, you'd better hope your heart is totally free of problems and your arteries are relatively clear of obstructions. Otherwise you're a candidate for a heart attack or some other serious cardiovascular malfunction. As an older person, you may have traces of heart disease even if you don't know it. That's why it's especially imperative that people over 40 cool down their body and heart gradually with a slow walk after any strenuous exercise session.

Ceasing exercise abruptly is bad enough. You compound the felony by . . .

• sitting or lying down immediately. Don't do it for at least 5 minutes after your exercise session ends.

• taking an immediate hot shower, plunging into a hot tub, or stepping into a sauna or steam room. However tempting and pleasant they may feel right after exercise, they're dangerous at that time. Stay away from them for 45 minutes to an hour. If you must shower soon after exercise, make sure the water is cool or lukewarm.

FREQUENCY, DURATION, AND INTENSITY

There are three concepts that exercise physiologists use to design exercise programs and predict their effectiveness. These are the frequency, duration, and intensity of a person's exercise effort.

Frequency refers to how often you exercise. By the end of our program, expect to be exercising aerobically about five days a week and performing strength training about three days a week. Infrequent sessions—only one or two every week—will do little to get you back in shape. At the opposite extreme, exercise that exceeds our program's target amounts isn't a good idea either. A six- or seven-day-a-week exercise regimen confers little additional cardiovascular or muscular benefit—while it actually increases the risk of injury.

Duration is the length of time devoted to each exercise session. Frequency and duration dovetail. For example, on our program, we start you out with shorter but more frequent aerobic sessions. Gradually these shorter workouts meld into one, less frequent, longer session. The maximum length of a daily aerobics session in our BioAction Plans is 50 minutes, and you do it five days a week.

In our BioAction Plans, we always specify the duration of

aerobics, However, keep in mind *our duration figure does not include a 5- to 10-minute, low-intensity aerobic warm-up in the beginning, followed by 5 minutes of stretching. Nor does it include a 5- to 10-minute aerobic cool-down at the end.* We would also like to see you end each aerobic and muscular conditioning workout with 5 minutes of stretching, though that's optional.

In our BioActions Plans, you will not see many duration guidelines for strength training. We leave it to your discretion to tailor the exercises to your particular needs. We let you decide how much rest to take between these exercises. For the average person, however, we recommend a rest of: ✓ 1 to 2 seconds between repetitions . . . ✓ 1 to 2 minutes between sets of repetitions . . . and ✓ 2 to 3 minutes between different exercises. However, the exact timing isn't very important. (Each lift, or muscle contraction, is known as a "repetition." A series of them is called a "set.")

The third concept is *intensity*. Intensity is a measurement of the level of your exertion during each workout. Technically, intensity means something slightly different within the context of the two different forms of exercise you'll be doing, but the distinction is really not worth going into here.

MEASURING THE INTENSITY OF YOUR *AEROBIC* EXERTION

In the world of exercise physiology, there are a few notions that enjoy a wide currency. One is an age-dependent formula that most exercise physiologists use to predict maximum aerobic capacity during peak exertion:

220 minus your age (in years) = Your Maximum Heart Rate

While this age-dependent, aerobic-capacity formula is much bandied about, we have some quarrel with it because it assumes that a person's age in years is the key variable. Using sophisticated equipment in our Tufts physiology lab, we've tested the aerobic capacity of hundreds of senior citizens. We're here to tell you that *maximal heart rate even in older people is remarkably variable.* In a healthy 55-year-old man, it may be 150; in another man of the same age, it could be 168. It's for this reason that we will not use this formula to tell you what intensity to exercise at. Since we can't actually measure your maximum heart rate, we don't want to run the risk of over- or underestimating your proper training intensity. What we'll use instead is the "Borg Perceived Exertion Scale" developed back in the early 1950s by the Swedish exercise physiologist Gunnar Borg. Today, this so-called Borg Scale is still very popular. It's used by scientists and physicians the world over to help them evaluate how hard people are exerting themselves on a treadmill or while undergoing some other medical test that requires exertion.

The scale, which we've duplicated below, helps people score their exercise effort on a scale of 6 to 20. A 6 rating is equivalent to no exertion. A rating of 20, on the other hand, represents a supreme effort. It means a person has pushed him- or herself to the absolute limit of endurance.

The Borg Scale figures prominently in both our 16-week and 12-week exercise programs in following chapters. In both these BioAction Plans, we give you a Borg Scale goal for your aerobic exercise and then ask you to evaluate whether you met that goal or not.

You may want to photocopy this scale and put it in a prominent place in your log book. If you work out at a health club, you might even want to tape it to the inside of your locker door so you have to see it while you're dressing for action.

Borg Scale of Perceived Exertion	
Numeric Rating of Your Exertion	**Verbal Description of Your Exertion**
6	None
7	Very, very light
8	
9	Very light
10	
11	Fairly light
12	
13	Somewhat hard
14	
15	Hard
16	
17	Very hard
18	
19	Very, very hard
20	

Source: G. A. Borg, *Medicine and Science in Sports and Exercise* 14 (1982): 377–87.

Before we leave this subject, we'd like to tell you a story that underscores the importance of monitoring your aerobic exercise intensity—especially if weight loss is one of your goals. We'll never forget the case of a woman who was involved in a weight-reducing program offered at a local hospital's clinic. She was obese; 46 percent of her body was fat. Her doctor had her on a 1,000-calorie-a-day diet, but the pounds still persisted. Finally her doctor prescribed exercise, which we would have done right off. Mysteriously, even exercise was having no effect.

When the woman told us she thought something was wrong because she never sweated during her exercise sessions, we got suspicious. We watched her one day as she pedaled away on a stationary exercise bike. No wonder the exercise was having no impact. She was "freewheeling"; she put no resistance on the pedals. A four-year-old could do it. In short, the intensity of her exercise was not high enough to account for any significant calorie expenditure. This is why our two BioAction Plans in Chapters 5 and 6 indicate intensity levels for all exercise we prescribe.

MEASURING THE INTENSITY OF YOUR *STRENGTH-BUILDING* EFFORT

We suspect that to many of you, muscular conditioning is an alien concept. You've never given it much thought before you picked up this book, and you've certainly never done it.

One of the most enduring myths of aging is that as we grow older, we lose the capacity to enlarge and strengthen our muscles. This myth has been aided and abetted by research studies designed by scientists whose expectations for their older test subjects was low. These researchers did not push their older subjects. Rather, they had them exercising at very minimal intensities. Indeed, most of the studies reported in the scientific literature describe older men and women lifting weights that are only 20 to 30 percent of the maximum amount of weight they can lift. We know that young people who train at such low intensities realize little increase in strength, so why should we assume that the situation with older people is any different?

At our studies at Tufts, we exercise our subjects at 80 percent of their one repetition maximum because we recognize that *to build strength appreciably a person must work out at that level*.

Let's review the concept of "one repetition maximum," or 1RM, discussed previously in Chapter 3. Exercise scientists express the intensity of muscular conditioning—also known as "resistance training"—in terms of 1RM. *One RM is the most weight a person can lift with one single movement or muscle contraction. It's a weight so heavy that the person cannot lift it again without resting for a while.* Clearly, your 1RM is like a fingerprint. It's unique to you.

In our BioAction Plans in Chapter 5 and 6, you'll generally be lifting a weight that's 80 percent of your 1RM. However, over the weeks of the program, you'll find that your 1RM—the heaviest weight you can lift at any given time—changes. As your muscles grow bigger and stronger, you'll be able to substitute heavier and heavier weights. That's why ours is a *progressive* resistance-training program.

Strength-Building Exercises

While you're following our BioAction Plan, these are the strength-building exercises you'll be doing throughout. However, once you graduate from the program, you'll be an experienced strength builder and will no doubt want to experiment with other routines you find in magazines or exercise books.

How many times and at what intensity will you be doing each exercise?

- In our BioAction Plans, you'll typically be performing two or three sets, each set consisting of 8 to 12 repetitions (lifts) ★ at the very most. This won't change much throughout the course of the program. What will change is the intensity or amount of weight that you lift.

- Every two weeks you'll be using the simple Strength Test for the upper and lower body (Chapter 3) to determine the amount of your upward weight adjustment. Once you know your new 1RM amount, it's simple arithmetic to calculate 80 percent of it. This is the amount of weight you'll be lifting over the course of the next two weeks.

★ Each lift, or muscle contraction, is known as a "repetition." A series of them is called a "set."

Our strength-building program has two targets: (1) the muscles of your upper body—those in your arms and shoulders; and (2) the muscles in your hips and legs, your lower body. Generally we'll be asking you to exercise the upper-body muscles on one day, followed by the lower-body muscles on the next.

Upper-Body Muscular Conditioning

• **The Easy Push-Up (or the Bent-Knee Push-Up).** Lie, face down, on the floor. Your feet should be together and the palms of your hands flat on the floor at either side of your chest as shown in the first drawing below. Support the weight of your upper body on your arms as you raise your body to the second position. Keep your back as flat as you can. This is a much easier push-up than the straight-knee version.

Throughout this routine, it's important to keep your breathing even. Inhale on the way up and exhale as you lower yourself gently back to the floor. Establish a steady rhythm, either by counting aloud or exercising to the strains of slow triumphal march music.

Keep in mind that the point of a push-up is to strengthen the muscles of your arms, chest, shoulders, and back. Doing this bent-knee push-up regularly, you'll eventually develop the strength to do the much harder straight-knee version.

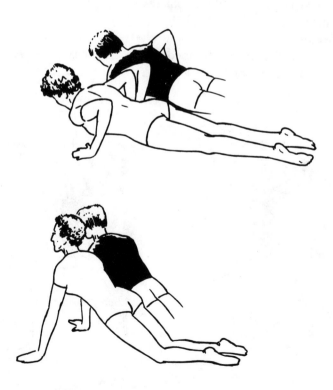

• **Arm Curl (or Biceps Curl).** Sit upright on a chair with your legs open and your feet flat on the floor. Your hand is holding a weight (or you're wearing a wrist weight as shown in the drawing). Flex your biceps muscle as you raise the weighted forearm. Bend your elbow and "curl" the weight to shoulder level. Slowly return your arm to the starting position.

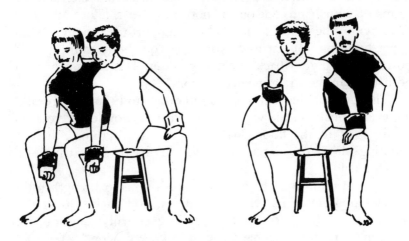

• **Chest (and Shoulder) Exercise.** Sit upright on a chair with one arm at your side, your hand holding a weight. Raise your arm slowly forward and up. Your elbow should be straight. Stop when your arm is fully extended above your head. *Slowly* return to the starting position.

• **Chest (and Shoulder) Exercise** (not illustrated). The last exercise has a variation that you may also want to try:

Lie on your back with your knees bent and your feet and lower back flat against the floor. Your arms should be spread-eagled on either side of your body, and each hand should be holding a weight.

Begin the exercise with your arms, elbows just slightly bent, elevated a few inches from the floor. Raise your arms slowly, keeping your elbows bent, until the weights meet above your body. Lower the weights slowly in the same arc.

When you do this exercise, it's important to keep the motion ever so slow, so that you won't injure yourself.

• **Upper-Arm (Triceps) Exercise.** Sit on a straight-backed chair. Hold a weight in each hand and raise both arms over your head. Bend one elbow as you lower one hand behind your head. Raise it back overhead, joining the other hand. Now, do the same thing with your other arm. Keep alternating.

Lower-Body Muscular Conditioning

• **Knee Extension (and Flexion).** (Drawing on next page.) Sit on a chair and tie or strap a weight to the ankle of one leg. Extend your leg so that it's as straight out in front of you as possible. Lower your leg back to its starting position. Rest no more than one second before repeating the same lift. Repeat the process with your other leg.

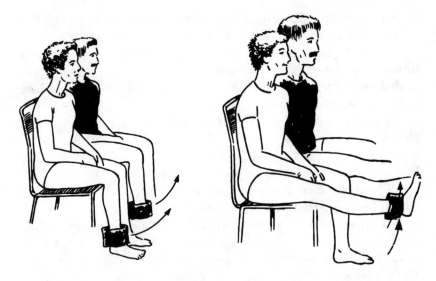

• **Hip and Knee Extension.** You may or may not need the back of a chair or the edge of a counter for support when you do this exercise.

Stand up straight, your feet planted firmly on the ground and your toes pointing outward. From this position bend your knees slightly, directing your body weight over your toes. *Do not do a deep-knee bend,* however. Keep your heels on the floor, as the benefits from this exercise occur with only a partial dip. Return to your starting, upright posture. Repeat.

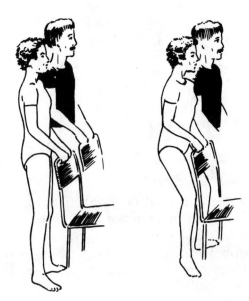

• **Step-Ups—or Stool Stepping.** This time you're going to step up and down from a stool using only one leg at a time. For example, if you start with your right leg, slowly raise your whole body onto the stool without using your left leg at all to help. Then, lower your whole body to the floor, once again relying only on the strength of your right leg muscles. Repeat this exercise 10 to 15 times. After a short rest, follow the same procedure with your left leg.

As your leg muscles grow stronger, you may want to make this exercise more challenging by adding weight. Don a weight belt or hold the weights you use for upper-body strength conditioning.

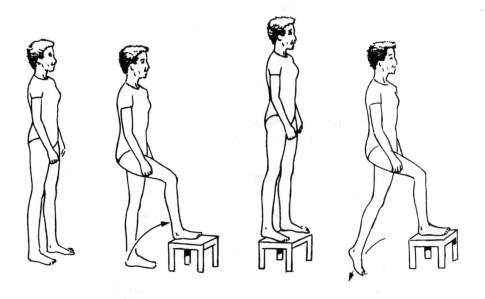

Strength-Building Do's and Don'ts

Like any form of exercise, resistance training can be done the right way—or the wrong way. To make sure you do it right, here are some pointers. (You should photocopy the Do's and Don'ts on pages 126 and 127 and put them in a prominent place where you can refer to them easily.)

Muscular Conditioning Pointers

Do

- concentrate, at least initially, on only a few key muscle groups. The exercises we illustrate in this chapter target those muscles in the shoulders, arms, and thighs that are most important for improving your normal, everyday activities. If you want to move beyond these muscle groups, wait until later, after you've completed our program.

- alternate muscle groups that you exercise. For strength-building purposes, we divide the body's muscles into upper and lower. Exercise the upper-body muscles one session and the lower-body muscles during the next.

- begin each session with several repetitions without a weight. This helps stretch that muscle group.

- perform all strength-building movements slowly, making sure the targeted joint has been put through its entire range of motion.

- breathe properly. *With each repetition, inhale before you lift, exhale as you lift, and inhale again as you slowly lower the weight to its beginning position.*

- take your time. Each repetition should take 6 to 9 seconds to complete. Rest for a few seconds between each lift.

- retest your strength periodically so you can readjust the amount of weight you should be lifting. On our program, we ask you to retest your strength once every two weeks using the Strength Test in Chapter 3.

- stay with the appropriate weight. *The appropriate amount of weight for you is the amount you can lift between 8 and 12 times, after which your muscles are too tired to continue.*

Don't

- swing a weight fast or bounce at the end of the movement.

- attempt to speed up a workout. It's dangerous. Even if you don't injure yourself, you'll find you're unduly fatigued at the end.

- hold your breath during a repetition. Holding the breath during weight lifting is known as the "Valsalva maneuver." It increases pressure in your chest, which can impair the normal flow of blood though the heart. End result: you feel dizzy or faint. Breathe with a normal rhythm, as we outlined above.

- exercise the same muscle groups more than once a day. You must allow your muscles sufficient time to recover.

- substitute lighter for heavier weights and simply do more repetitions.

- substitute heavier weights than we specify, thinking you can just do fewer repetitions and speed up your workout.

Source: Maria A. Fiatarone, M.D., "Keep Moving, Fitness After Fifty" program

Eventually, after you've built your endurance and gotten your body back in pretty good shape, you may want to combine aerobics and strength training into one long daily session. In many ways, it's the safest and most efficient way to proceed. Why? For one thing, you have to do only one warm-up and cool-down and one set of stretches at the beginning and end. Moreover, by the time you get to the weight-training part of the session, your tendons, ligaments, and muscles will be fully warmed up and less likely to become pulled. In short, it lowers your chances of an exercise-induced injury.

A final safety note:

Don't confuse isotonic exercise (weight lifting and stretching) with isometric exercise. This latter type of exercise involves pushing against an immovable object, like a wall. During an isometric contraction, the critical flow of blood to the muscles stops, reducing the effectiveness of that exercise. In addition, people commonly hold their breath during an isometric maneuver. This can cause a rise in blood pressure that's not healthy. It's particularly dangerous for people with heart conditions—and many older people are suffering from heart disease but don't know it.

In contrast, isotonic exercise is much safer. Lifting a weight while moving a joint through its full range of motion allows blood to continue flowing to the working muscles. This builds muscular strength while improving flexibility.

TRACKING YOUR PROGRESS: THAT ALL-IMPORTANT LOG BOOK

There's one thing we've learned over the years about people and exercise. Structure and formality are important, especially in those tentative early days when novitiates are thinking about quitting at least as much as they're thinking about persevering.

We've tried to make our instructions more visually appealing—and easier to follow—by laying them out in the form of an exercise log book. In Chapters 5 and 6, you'll find our log book pages filled with instructions, although we also leave room in key places for you to fill in the blanks concerning your aerobic intensity or your strength-training weight and repetition goals.

Glance at the empty log book page, opposite, for a moment. Notice there's room for you to write in your dietary goal(s) for the week. After you read Part Four, which covers nutrition, you'll no doubt have some eating objectives. Here's where you can keep track of them—and note your progress in achieving them. Underneath are three columns headed "Aerobics Exercise Track," "Strength-Building Track," and "Comments." As you're learning in this book, there's an important distinction between aerobic exercise and strength training. It's important to make that distinction in your mind as well as on paper.

ESSENTIAL EXERCISE EQUIPMENT

For novice exercisers, here's a run-down of the attire and equipment you'll need to assemble before you start. It's minimal:

Since walking is the aerobic exercise we strongly recommend, especially for the low-fit, you need a sturdy pair of walking shoes. A good running shoe is another possibility. Just make sure the shoes you choose have a sturdy sole, provide adequate support, and are comfortable. You'll be putting a lot of miles on these shoes, so choose carefully. The last thing you need is aching feet.

A three-ring binder is *de rigueur*. You'll turn this into your exercise log book, filled with photocopied sections of this book. In it you'll put your BioAction Plan, Borg Intensity Scale, our point system scoring rules, as well as salient instructions—strength-training do's and don'ts, stretching tips, etc.

A special exercise wardrobe shouldn't be necessary. Loose-fitting clothing that affords ease of movement is all that's required. You probably have closets and drawers packed with them already.

When walking during the summer, wear a wide-brimmed hat to shade you from the sun's rays and wear lighter clothing. Shorts and a T-shirt are ideal. Remember, during exercise your body generates a lot of heat. That heat causes sweat to form on your

Figure 4-2

My nutritional goals are . . .

■
■

WEEK OF _____

	Aerobic Exercise Track		Strength-Building Track	Comments
	Frequency & Duration	Borg Scale Intensity *	Frequency & Duration	80% of Maximal Lifting Capacity †	
Day 1:		*(Fill in, please.)*		*(Fill in, please.)*	
Day 2:		*(Fill in, please.)*		*(Fill in, please.)*	
Day 3:		*(Fill in, please.)*		*(Fill in, please.)*	
Day 4:		*(Fill in, please.)*		*(Fill in, please.)*	
Day 5:		*(Fill in, please.)*		*(Fill in, please.)*	
Day 6:		*(Fill in, please.)*		*(Fill in, please.)*	
Day 7:		*(Fill in, please.)*		*(Fill in, please.)*	

* We explained the concept of intensity and reproduced the Borg Scale earlier in this chapter.

† Maximal lifting capacity, which is the most weight you can lift with one try, is predicted by how many times you lift a weight of X amount. Chapter 3's upper- and lower-body Strength Test is used to establish maximal lifting capacity. You should retake the Strength Test every two weeks to readjust the amount of weight you lift. To decide the weight amount, take 80% of your maximal lifting capacity. For example, if your maximal lifting capacity is 25 pounds, you should lift 20 pounds during the upcoming two weeks. If you're faithfully adhering to the program, your readjustment should always be upward.

129

skin. Evaporating sweat is the way your body dissipates heat and cools itself.

You should never attempt to lose weight by entrapping your sweat inside rubberized or non-breathable-fabric clothing. This is dangerous and doesn't make you shed fat anyway. All you lose is body fluids.

When walking during cooler weather, appropriate attire is loose, layered clothing. The air pockets between your clothing and your skin and between each layer of clothing act as insulation. Be especially careful to protect your fingers and toes. Wear gloves and either heavier socks or two pairs of socks. While exercise tends to increase the flow of blood to these extremities, this is not the case with everyone. If you always feel the cold in your fingers and toes, you're susceptible to frostbite and need to take extra precautions.

For the strength-building sessions, of course, you'll need weights. Contrary to what you might expect, our muscular conditioning program does not require a lot of equipment and claptrap. In fact, homemade weights are fine if you're determined not to invest a penny in any special gear.

You can make a weight of varying poundage by filling empty plastic containers or milk cartons with sand or water. A sturdy plastic bag filled with dirt can also serve as a weight for knee extensions. To make heavier weights, use lead shot. Weigh your homemade weights on your bathroom scale. For safety's sake, we would advise you to use containers with handles for all upper-body workouts. They're easier to grasp.

Commercial weights—called "free weights" or "hand weights"—are ideal for upper-body workouts. For lower-body strength training, weight cuffs or spats are useful. We've known people who wear them to work, tucked around their ankles under their pants, to build strength in their leg muscles and increase their daily calorie expenditure.

If you're going to buy weights, you'll need a variety great enough to accommodate the fact that you'll be increasing the amount of weight that you lift every two weeks. As a faithful adherent of our program, expect up to a 150 percent increase in the amount of weight you can lift over the full course of the program. That means that a person whose initial test weight is 25 pounds should purchase weights in 5- or 10-pound increments, up to a top weight of 65 pounds.

Believe it or not, a special kind of rubber band of stretchy latex, with the brand name Dyna-Band,★ are also "weights" of a sort. When you stretch them, they function like weights, causing a targeted set of muscles to contract. We strongly recommend them. They enable you to exercise a wider variety of muscles than weight lifting generally does, they're inexpensive, and they're portable (for those of you who travel a lot). In addition, they're fun, easy to use, and come in a variety of different thicknesses and resistances. (Resistance is indicated by a color code—the darker the color of the band, the greater its resistance.)

A key item for serious exercisers is a weight belt. It's worn around the waist and helps both to build strength and to increase the intensity of an aerobic session.

In our BioAction Plan for the moderately fit, we strongly recommend wearing a weight belt at times other than just during formal workouts. Since there's a limit to how fast a person can walk, a weight belt becomes a way to increase the intensity without jogging or running. This extra weight carried around the midsection also helps burn off more calories, if weight loss is a goal.

Some Optional Exercise Gear

We outlined the bare essentials above. Here are a few not-so-essential items you might consider:

You may be tempted to invest in a heart rate monitor to avoid the delicate process of taking your pulse after exertion. Unfortunately, the cheaper ones are close to useless. They're erratic and inaccurate. You'll have to spend close to $300 to get one that's accurate enough to be worthwhile.

Stationary bicycles—called "cycle ergometers"—have their advantages, to be sure. They're easy to use, they're indispensable for people who have trouble walking, and they're the best aerobic alternative to walking in inclement weather. We realize that stationary bikes are steady sellers in the sporting goods stores, but, in our experience, most are all but abandoned after the initial pride of ownership wears off.

We'll wager there are more unused stationary bikes in America than used cars! Why? Because riding on a stationary bike is B-O-R-I-N-G. About the only way to make it palatable is to watch TV while you're doing it.

★ You can order a set of Dyna-Bands, accompanied by an illustrated pamphlet depicting specific exercises you can do with them, from Fitness Wholesale, Inc., 3064 West Edgerton, Silver Lake, OH 44224 (1-800-537-5512).

We frequently use stationary bikes in our research, it's true. For us, the advantage is that the intensity of a person's exercise can be precisely set. But we've noticed that most of our volunteers who exercise on bikes at our research center do not continue exercising once they depart. In contrast, people who have participated in our walking programs to measure and improve fitness usually continue exercising on their own long after they've left our formal program.

However, don't let us dissuade you from buying a bike if you're sure it's your cup of tea. We recommend the following brands:

Monark bikes, originally made in Sweden, are the Cadillacs of the ergometer world and what we use in our physiology laboratory at Tufts. For our purposes, their best feature is accuracy. The workload intensity you set is what you get; there's little drift.

Tunturi bikes are also dependable and durable. They may not be quite as accurate as the Monark, but that shouldn't matter to you.

SOME SAFETY ADVICE ABOUT
EXERTION AND DEHYDRATION

As you learned when we discussed Biomarker #10 concerning the body's internal temperature control back in Chapter 2, the older you are, the more careful you must be about replacing fluids lost through exercise. Because older people have a reduced sensation of thirst, to be really safe, you should make a habit of weighing yourself before and after exercise. Then, drink back in fluids the amount of weight that you lost during exertion, because all of this sudden weight loss is water.

For example, if your weight dropped 1½ pounds (0.68 kilograms) as a result of your exercise, this would mean that you lost a little more than ½ quart of water. As the chart opposite shows, ½ quart of water weighs almost one pound. This chart will make it easy for you to judge how much liquid—that's a *nonalcoholic* liquid or beverage—you should drink to replace lost body fluids.

Not only should you avoid alcohol directly after exercise, we warn you not to drink it beforehand, either. Alcohol exacerbates body-fluid depletion by making you urinate more.

Incidentally, if your urine is dark, it could mean that you're dehydrated. In such an instance, drink enough fluid to turn your urine clear.

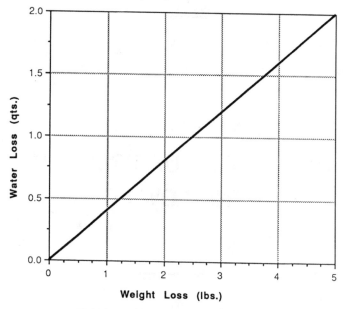

WATER-WEIGHT LOSS CHART

Here's how to use this chart:

Weigh yourself before and after you exercise. Along the bottom horizontal axis of the chart, place a dot corresponding to the amount of weight you lost during exertion. With a ruler, trace a line vertically up from this dot to the diagonal line. From this spot trace a horizontal line left to meet the chart's left edge. This intersection tells you how much fluid to drink.

LOSING WEIGHT ON AN EXERCISE PROGRAM

Keep this fact in mind: One pound of body fat is equivalent to 3,500 kilocalories (kcals), what most people simply call "calories." Discounting the first few weeks on our BioAction Plans, which are just preliminary, we have designed our program so that body-fat loss is a real possibility. You should be able to lose about one pound of body fat every two weeks *provided* your dietary intake of calories doesn't go up because your increased activity is stimulating your appetite and you're eating more. Of course, you can accelerate your body-fat loss somewhat over the course of the program by sharply reducing the fat, especially saturated fat, in your diet. We recommend relatively slow body-fat loss because slow loss is generally more permanent and is the safest route to your ultimate goal—which is a more muscular body with less fat tissue.

CHAPTER
5

BIOACTION PLAN A: 16-WEEK CATCH-UP PROGRAM FOR THE LOW-FIT

Over the past six years, we've had hundreds of older men and women pass through the doors of our laboratory to begin an exercise program with the hope "it will give me more energy." Betty, who was 62 years old when she volunteered for our program, was one of them.

Betty complained constantly about fatigue. "I don't understand it," she told us. "I'm doing less than ever now, but I feel weary all the time."

She claimed a shopping trip to the local mall, where she'd be walking around without sitting down for several hours at a stretch, would leave her exhausted. Ordinary housework and gardening chores were a challenge. By seven o'clock every evening, all she wanted to do was relax and watch TV. There was no way she had any energy left over to go out socially with her husband, who was full of pep and always ready for a new adventure.

Betty thought she might be deficient in vitamins or iron. "Maybe I've got 'tired blood,' " she joked, recalling the famous iron tonic commercial from years ago.

Betty was deficient—in *physical* activity!

Betty's problem is not uncommon, especially among retired people. Like most people, Betty had looked forward to retirement for years. She viewed it as a time of rest and relaxation, no more hectic work schedules or pressure to produce on a tight timetable. Unfortunately, she, like so many people before her, was discovering her newfound idleness had a dark side. It seemed to translate into chronic fatigue.

As a volunteer at our center, Betty underwent the usual health evaluation. She wasn't the least bit anemic; a blood analysis showed no nutrient deficiencies. Everything was normal except for her extremely low fitness level.

To put it in the terminology of this book, Betty's biomarkers were showing. Among other things, she'd gained 15 pounds over the previous two years, making her moderately overweight. Due to her sedentary lifestyle and complete lack of exercise, she had a disproportionate amount of fat to lean-body mass, even for her age. Her body was 50 percent fat! The average for her age is about 42 percent fat. Walking at a brisk 3½-mile-per-hour pace for 15 minutes placed her at almost 100 percent of her maximal aerobic capacity—an elevated level that people cannot sustain for more than a couple of minutes.

Betty saw no correlation among her weight, her lack of conditioning, and her enervated state. Most people don't. We fear it will take several more books like this one before the average person recognizes the strong link that exists between fitness levels and energy levels.

While our initial testing put Betty in the category of low fitness, all was not doom and gloom for her. In fact, Betty—and any of you who scored in the low-fitness group on Chapter 3's tests—are in the enviable position of having more to look forward to than those lively folks who are already in pretty good shape. This may seem like a contradiction. It's not.

Research shows that the single best predictor of how much gain an individual will make during an exercise program is his or her baseline fitness. This stands to reason, for . . . *the lower the starting level, the greater the potential gain.*

As an armchair idler—or a "couch potato," in the parlance of the day—you probably feel embarrassed by your mediocre physical condition and intimidated by the people your age who exercise regularly and exude good health. You shouldn't be—and don't need to be. By following the BioAction Plan we've laid out for you in this chapter, you can be in their shoes in a matter of a few short months.

Betty was. But it took about five weeks before she began to see light at the end of each exercise session.

Betty's biggest fear in starting this plan was that it would wear her out, making her even more tired than usual at the end of each day. For the first week or two she didn't notice any change for the better in the way she felt. Realizing that she'd be increasing the amount of exercise she did each week, she became quite convinced that her worst fears would come to pass. However, by the fifth week, instead of fatigue at the end of an exercise session, she felt the warm afterglow of accomplishment. Instead of being exhausted each evening, she was more energized.

By the fourteenth week, the changes in Betty were readily apparent to all who'd known her before she started. She'd lost eight pounds and was enthusiastically involving herself in more activities than she'd undertaken in years. At the start of the plan, Betty had resented the idea of devoting so much time to exercise. Later she realized that exercise was stretching her daily complement of time and energy. Because of exercise, she actually had more time each day to devote to new projects.

The new Betty was an inspiration to several of her friends who had turned Betty down weeks earlier when she'd asked them to be her BioAction Plan comrades. They joined the program, too.

If this sounds like pie-in-the-sky to you, let us assure you that Betty's experience is not unique. Given your sedentary state, it's important to realize that *anything at all that you do over and above what you're doing already will help pull you out of your lethargy.* Even without embarking on an exercise campaign, *you can do a lot to raise your fitness level just by making the right tiny choices every day.*

What are we talking about?

Take the small, unconscious choice you make every morning when you step out of your fifth-floor apartment to exit the building. Naturally, you wait in front of the elevator for it to come. But is a vertical trip via something invented back at the turn of the century really all that natural? Indeed, where it is written that you have to take an elevator, either to go down or up within your building? Ask yourself, Is this my only option? Consider, if you will, walking down those five flights. For sure, this small bit of exertion would be an excellent way to start injecting more motion into the routine of your daily life.

We'll wager there are twenty or more minor-league choices just like this one you make every day. Ask yourself: Why should I automatically hop in my car to pick up a newspaper eight blocks away? Or take the matter of grocery shopping. We're referring not to those times when you need a week's worth of food, but to the grocery errands that involve only a few staples like milk and bread. Stop relying on wheels all the time. Rely, instead, on your own two legs. Carry that bag of groceries those three blocks back to your house, amble along and admire the scenery along the way.

We urge you to consider buying a bicycle to use for both necessary locomotion and pure enjoyment. Get your friends to buy bikes too and accompany you on lengthy, fun-filled rides.

And speaking of friends and family, above all, don't let their jeers curb your newfound ardor for activity. Be prepared when they call you loony. Instead of caving in to their bad counsel and influence, turn the tables and convert them to your new way of thinking—your new, more sensible, *and natural* approach to life.

As we have emphasized many times, friends with whom you share the Biomarkers Program are a key to your success. Starting an exercise training program is like entering a brave new world. Doing it alone is scary. With no friend or spouse at your side, you'll get no reinforcement from people whose opinions you value.

A WALKING TEST TO GET YOU STARTED

One of the best ways to keep people motivated is to give them ways to measure their progress. With this in mind, we want to start you off with a very simple walking test.

Quarter-Mile Walking Test

- Mark off a quarter-mile course using landmarks—telephone poles, fire hydrants, a specific store—as your guide. The distance does not have to be exact as long as you always perform this test in the same place so you're always comparing apples and apples, not apples and oranges.

- Walk the course *as rapidly as you can,* timing yourself using a watch with a second hand or a stopwatch. At the end of the course, stop dead in your tracks and immediately measure your pulse rate for 30 seconds—or have your exercise partner measure it for you. (See instructions in Chapter 3, figure 3–1.)

Throughout the 16 weeks of the BioAction Plan that follows, ideally you should repeat this little self-test once every two weeks, during weeks 3, 5, 7, 9, 11, 13, and 15. Each time, walk the course as fast as you can and monitor your pulse for half a minute. As you become better trained, this test will become easier and easier —and your pulse rate will be lower and lower. That's a good sign. Conditioned people are people with efficient cardiovascular systems. It takes progressively more exertion to make their hearts pump hard.

Over the course of the 16 weeks, you'll also be retesting your strength—using the self-test in Chapter 3—every two or three weeks—during weeks 5, 7, 10, 12, 14, and 16.

A WORD ABOUT SORENESS

Before you begin, we must caution you about a phenomenon termed "delayed onset muscle soreness." This refers to those proverbial aches and pains that we all feel a day or two after we use muscles that usually remain dormant. The pain we feel is symptomatic. Our bodies are alerting us to microscopic tears in our muscle cells. The tears are causing inflammation and a slight swelling, hence the pain.

While this sounds serious, it isn't. It's completely natural. Indeed, this minor damage seems to be right and necessary. It appears to be the way our muscles adapt to exercise training.

We advise you *not to take aspirin, any over-the-counter painkiller, or ibuprofen for the soreness you're sure to feel as you begin the program.* New research indicates that these medications may, in fact, slow down the beneficial adaptations that training forces your muscles to make.[1]

We can't close this chapter without giving you a resounding pat on the back. You've achieved what most people only fantasize about. Unfortunately, it's still all too true that the average person in the industrialized nations of the world does *not* exercise. By finishing this program, even if your regular adherence to our

WEEK 1*

Day	Aerobic Exercise Track — Frequency & Duration	Borg Scale Intensity*	Strength-Building Track	Comments
Day 1:	Divide your exercise walk into two separate sessions of 10 minutes each. Total time: no more than 20 minutes.	You determine the intensity: *(Fill in, please.)*		If you have trouble walking for 10 minutes straight, divide your exercise into even smaller sessions of 5 or 6 minutes each.
Day 2:	Follow yesterday's regimen. Do the same concerning your exercise intensity.	*(Fill in, please.)*		Attempt to exercise at the same time every day. This way, you'll begin to incorporate your training into your daily routine.
Day 3:	Walk or cycle for a total of 20 minutes in two separate sessions.	Increase your pace slightly. If you're 12, aim for 13 or 14. *(Fill in, please.)*		Every time you exercise, get into the habit of evaluating the intensity of your exertion. As you become better trained and more conditioned, you'll find your Borg Scale number will remain constant even though your pace quickens.
Day 4:	Repeat: Walk or cycle for a total of 20 minutes in two separate sessions.	Aim for 13 or 14. *(Fill in, please.)*	There is no muscular conditioning during the first 12 days of BioAction Plan A.	
Day 5:	REST. If you feel restless and want to do something, just walk or cycle for one 10-minute session.			Exercise training is a slow, gradual process; and rest is a critical part of any good program.
Day 6:	Today, undertake two 15-minute sessions—or three 10-minute ones.	Moderate: 13 or 14. *(Fill in, please.)*		Some of your Biomarkers have already started to change. For example, there are already measurable changes in your glucose tolerance; and your muscles have more carbohydrate stores than they did before Day 1.
Day 7:	Once again, either walk or cycle for two 15-minute sessions—or three 10-minute ones.	Shoot for 14+. *(Fill in, please.)*		You've finished your first week! The hardest part, believe it or not, is already over.

* We explained the concept of intensity and reproduced the Borg Scale in Chapter 4.

	Aerobic Exercise Track Frequency & Duration	Borg Scale Intensity	Strength-Building Track	Comments
Day 8:	Your second week begins with two 15-minute walking and/or cycling workouts.	Your target: 14. *(Fill in, please.)*		At the end of each workout, continue walking at a slower pace—11 on the Borg Scale—for 5 minutes longer, then stop and stretch. You should be exercising on a regular schedule. Being hit or miss about it—exercising at different times each day—is a recipe for failure, at least in the beginning when your commitment may tend to waver.
Day 9:	Let's up the ante to two 20-minute sessions—walking or cycling	Your target: 14. *(Fill in, please.)*		Don't forget to warm up for 5 minutes and cool down for 10. And never, *ever* forget to s-t-r-e-t-c-h, one of the main safeguards against exercise injury.
Day 10:	Free day. No formal exercise required.			The fact that there is no formal exercise today doesn't mean immobility. Walk up some stairs instead of taking an elevator. Cycle over to a friend's house across town instead of hopping in the car. In the grocery store, carry a basket instead of pushing a cart. In general, stop looking for ways to avoid movement and activity.
Day 11:	For 30 consecutive minutes, either walk or take a bike ride.	Less intense: 13. *(Fill in, please.)*		Simply slow down if you find you have trouble completing 30 minutes at this pace. Duration is more important than intensity.
Day 12:	Once again, a workout of 30 minutes—without rest—is your goal.	Goal: To maintain a steady pace. *(Fill in, please.)*		Exercise should already be easier for you than it was on the first couple of days. This is because your muscles immediately began adapting to the regular bout of daily exercise. The muscles you've been using are now beginning to take up and process oxygen more efficiently; and your aerobic fitness (VO_2 max.) has already started to improve.

WEEK 2 (continued)

	Aerobic Exercise Track		Strength–Building Track		Comments
	Frequency & Duration	Borg Scale Intensity	Frequency & Duration	% of Maximal Lifting Capacity*	
Day 13:			Warm up as we described in Chapter 4; it should include at least 5 minutes of stretching. Then for the next 20 minutes or so, build your strength. The exercises below are illustrated in Chapter 4. ■ Knee extension: 3 sets of 8–10 repetitions for each leg. ■ Hip and knee extension: 3 sets of 8–10 repetitions. ■ Push-ups: 3 sets of 3; 6 if you can manage it. ■ Now work the biceps in your arms: 3 sets of 7 curls.	Your goal is to lift 80% of your maximal lifting capacity. You determined your maximal lifting capacity when you took the Strength Test in Chapter 3. _____ (Fill in, please.)	This is a banner day—*your first day of strength training!* As you grow stronger, you'll be lifting progressively more weight. It's very important to be lifting a weight that's right for you—and your strength. If your joints hurt while you're lifting, drop down to a lower weight. Don't forget to warm up, s-t-r-e-t-c-h, and cool down.
Day 14:	Walk or ride your bike for a minimum of 30 minutes.	Goal: 14 or 15. _____ (Fill in, please.)	Do the same routine as yesterday.	Goal: 14 or 15. _____ (Fill in, please.)	If you're up to it, extend your aerobic exercise period to a total of 40 minutes, but no more. It's likely your muscles are sore, but have no fear. The soreness will be gone in a few days.

* Maximal lifting capacity, which is the most weight you can lift with one try, is predicted by how many times you lift a weight of X amount. You established your maximal lifting capacity when you took the Strength Test in Chapter 3. To decide how much weight to lift each day, take the goal percentage (see amount listed in Strength-Building Track above) of your maximal lifting capacity. For example, on Day 13 the goal, as usual, is 80%. If your maximal lifting capacity is 25 pounds, you should lift 20 pounds.

You'll be retaking the Strength Test every 2 weeks and readjusting the amount of weight that you lift. If you're faithfully adhering to the program, your readjustment should always be upward.

WEEK 3

	Aerobic Exercise Track — Frequency & Duration	Borg Scale Intensity	Strength-Building Track — Frequency & Duration	% of Maximal Lifting Capacity	Comments
Day 15:	Walk or ride for 40 minutes without rest. Option: Retake the Quarter-Mile Walking Test.	Goal: 15 *(Fill in, please.)*	No, there's no strength training on the agenda today. Your muscles need time to recover.		Drink a glass of water before you begin your aerobic session. Get used to drinking plenty of fluids before and after each workout whether you feel thirsty or not.
Day 16:	One 40-minute walk or bicycle ride is your aerobics regimen today.	Goal: 13 or 14. *(Fill in, please.)*	Repeat Day 13's strength-training routine.	Goal: 80% _____ *(Fill in, please.)*	Test how your daily aerobics training has affected your heart rate. Retake the quarter-mile walking test we outlined in the beginning of this chapter and measure your heart-rate response immediately afterward. It may be too soon to notice any improvement. But when you take the test again in another two weeks, you should notice that your pulse is slower.
Day 17:	REST. If you've been faithful to our program, you deserve it.		No strength-building today.		
Day 18:	Forty minutes of aerobics, please.	Goal: 15. *(Fill in, please.)*	Again, no strength training.		Stretching becomes even more important now that you're incorporating strength training into your overall exercise effort.
Day 19:	Forty minutes of aerobics.	Goal: 15. *(Fill in, please.)*	Knee extensions: 3 sets of 8 repetitions. Hip & knee extension: 3 sets of 8 repetitions.	Aim for 80% of your maximal lifting capacity _____ *(Fill in, please.)*	There are two components of your strength-building program: upper- and lower-body training. We want you to alternate between the two. Today, it was your lower body that got the workout. Tomorrow, it will be your upper body.

WEEK 3 (continued)

	Aerobic Exercise Track		Strength-Building Track		Comments
	Frequency & Duration	Borg Scale Intensity	Frequency & Duration	% of Maximal Lifting Capacity	
Day 20:	Forty minutes of aerobics.	Goal: 15. *(Fill in, please.)*	Push-ups: 3 sets of 4–6 repetitions. Arm curl: 3 sets of 6 repetitions for each arm.	80% of your max. lifting capacity *(Fill in, please.)*	If you find the number of repetitions too tough for you, decrease the amount of weight. If you're having trouble doing the exercise even with little or no weights, take more time to recover between each repetition and set.
Day 21:	No aerobic exercise today. However, in its place, do at least 5 extra minutes of warm-up and stretching.		Knee extension & flexion: 3 sets of 8 repetitions. Hip & knee extension: 3 sets of 8 repetitions.	80% of your max. lifting capacity *(Fill in, please.)*	Those lead Biomarkers—your lean-body mass and strength—have started to respond by Day 21 of the program—that is, if you've been faithful about following our instructions. Here's what's occurring in your body: ■ You're already a little stronger, probably because your body is "learning" to use more and more muscle cells every time you lift and put stress on your muscles. ■ In turn, your muscles are undergoing a kind of renovation. They have begun the process of tearing down damaged muscle and replacing it with newly rebuilt protein tissue. ■ The capillaries around your muscle cells are growing to increase their blood and oxygen supply.

WEEK 4

	Aerobic Exercise Track		Strength-Building Track		Comments
	Frequency & Duration	Borg Scale Intensity	Frequency & Duration	% of Maximal Lifting Capacity	
Day 22:	Forty-five minutes of aerobics today.	Goal: 15 *(Fill in, please.)*	Bent-knee push-ups: 3 sets of 5 each. Arm curl: 3 sets of 9.	80% of your max. lifting capacity *(Fill in, please.)*	As you begin your fourth week of training, you should start to notice improvement in your ability to perform both the aerobics and strength-building exercises.
Day 23:	Another 45 minutes of aerobics	Goal: 16 *(Fill in, please.)*	No strength training today.		The longer and more intense your aerobic exercise becomes, the more important the cool-down phase after you finish becomes.

Day 24:	Once again, 45 minutes of aerobics.	Goal: 16
		
		(Fill in, please.)
			Knee extension & flexion:
			3 sets of 9 repetitions.
			Hip & knee extension:
			3 sets of 9 repetitions.
				80% of your max. capacity
				
				(Fill in, please.)
					We hope you aren't forgetting the very important stretching phase of the program.
					   Stretching is a long, slow process. Although you may not as yet feel any changes in flexibility, they're there. Every time you stretch, you're stimulating the muscles and tendons to lengthen.
					   Your maximal lifting capacity has also increased, in most cases, noticeably.

Day 25:	Forty-five minutes of aerobics.	Goal: 16 *(Fill in, please.)*	Push-ups: 3 sets of 5 each. Arm curl: 3 sets of 6.	80% of your max. capacity *(Fill in, please.)*	By now, you should be using up 200 to 350 extra calories per day. If you've been careful to consume less calories—especially fat calories—you should also be noticing a drop in your weight.

WEEK 4 (continued)

	Aerobic Exercise Track		Strength-Building Track		Comments
	Frequency & Duration	Borg Scale Intensity	Frequency & Duration	% of Maximal Lifting Capacity	
Day 26:	REST.		REST.		You deserve a reward for staying with the Biomarkers Program this long. Go out with your spouse, your family, or your friends. Plan an outing that requires a little walking—a trip to an amusement park or a museum, perhaps.
Day 27:	Back to that daily regimen of 45 minutes of aerobics.	Goal: 15 or 16 _____ _(Fill in, please.)_	Knee extension & flexion: 3 sets of 6 repetitions. Hip & knee extension: 3 sets of 8 repetitions.	80% of your max. capacity _____ _(Fill in, please.)_	Both the strength training and aerobic phases of the program are putting stress on your skeleton, so it too responds and starts to make positive adaptations. When you sweat now during your workout, you're not losing as much potassium, sodium, and other important electrolytes as before. Exercise stimulates the sweat glands to decrease the losses of these important substances. But keep drinking lots of water.
Day 28:	Forty-five minutes of aerobics.	Goal: 16 _____ _(Fill in, please.)_	Take a respite from strength training today.		

WEEK 5

	Aerobic Exercise Track		Strength-Building Track		Comments
	Frequency & Duration	Borg Scale Intensity	Frequency & Duration	% of Maximal Lifting Capacity	
Day 29:	The usual: 45 minutes of aerobics. Option: Retake the Quarter-Mile Walking Test.	Goal: 16 _(Fill in, please.)_	Test the improvement in your strength by repeating the tests you first took in Chapter 3. Use the tests' results to adjust the weight you'll lift over the next two weeks.		Chapter 3's tests will serve as the basis for the intensity of your strength training over the next two weeks—Weeks 5 and 6 of the program.
Day 30:	Only 30 minutes of aerobics.	Goal: 16 _(Fill in, please.)_	Push-ups: 3 sets of 7 repetitions. Arm curl: 3 sets of 7 repetitions. Chest & shoulder exercise: 2 sets of 5 repetitions.	Goal: 80% _(Fill in, please.)_	For the second time—you also did it at the beginning of Week 3—retake the quarter-mile walking test. What's your heart-rate response this time? If you're able to perform more than 15 push-ups, add some weight to your body, via a weight belt, to increase the intensity.
Day 31:	Back to 45 minutes of aerobics.	Goal: 16 _(Fill in, please.)_	Knee extension: 3 sets of 8 repetitions. Hip & knee extension: 3 sets of 8.	Goal: 80% _(Fill in, please.)_	As you increase the pace and intensity of your exercise, you're also increasing the burn-off of calories, enabling those of you who need to shed body fat to foster this process. You've used approximately 400 extra calories today. This is about equivalent to the calories you would put into your body by eating one slice of a fruit pie.
Day 32:	No aerobics today.		Push-ups: 3 sets of 8 repetitions. Chest & shoulder exercise: 3 sets of 8 repetitions.	Goal: 80% _(Fill in, please.)_	

WEEK 5 (continued)

	Aerobic Exercise Track		Strength-Building Track		Comments
	Frequency & Duration	Borg Scale Intensity	Frequency & Duration	% of Maximal Lifting Capacity	
Day 33:	Forty-five minutes of aerobics.	Goal: 16 _____ *(Fill in, please.)*	Leg exercises of your choice: 3 sets of 8 repetitions.	Goal: 80% _____ *(Fill in, please.)*	If you're feeling particularly strong, try a fourth set of 5 or 6 repetitions.
Day 34:	Same as yesterday and tomorrow.	Goal: 16 _____ *(Fill in, please.)*	No strength building today.		Make certain that you're drinking plenty of fluids. Force yourself, if you have to. As exercise causes you to sweat more, it's important to replace the water your body is losing.
Day 35:	Forty-five minutes of aerobics.	Goal: 16 _____ *(Fill in, please.)*	Push-ups: 3 sets of 7 repetitions. Chest, shoulder and upper arm exercises: 3 sets of 8 repetitions.	Goal: 80% _____ *(Fill in, please.)*	By now, you should be significantly stronger than when you started 5 weeks ago.

To continue tracking your progress on the program, you'll need to make photocopies of the empty log book in Chapter 4 (figure 4-2).

WEEK 6

Aerobic Exercise Track	Strength-Building Track	Comments
By now, you see the pattern. You should be walking or cycling 45 minutes per day, 4 to 5 days per week. Your intensity should be close to 16.	Alternate upper- and lower-body exercise from one day to the next. You should be doing 3 sets of repetitions of each exercise at about 80% of your maximal lifting capacity. * Perform these strength-training routines 5 or 6 times per week. No more, though. Your muscles need at least one day of rest.	Aerobics: By this time, your walking pace should be brisk. As an option, you may want to slow down your pace and wear a weighted belt (see Chapter 4). When you wear the belt, your intensity, despite the slower speed, remains the same. (Yes, hand and ankle weights are okay, too. However, weight on your back is easier to carry.) Strength Building: We urge you not to repeat an exercise routine on consecutive days, to allow time for each set of muscles to recover. Alternating should keep your muscles from feeling sore. If not, then you're probably lifting too much weight. We suggest you cut back.

* From here to the end of BioAction Plan A, your goal percentage will remain 80% of your maximal lifting capacity.

WEEK 7

Aerobic Exercise Track	Strength-Building Track	Comments

Repeat all the exercises of the previous week. By now, you and your partner(s) should be exercising on a very regular schedule.
 	
Option: Begin the week by retaking the Quarter-Mile Walking Test.
	The first day of this week, remeasure your muscle strength with the tests in Chapter 3 then readjust your weights.	Even if you cannot see that your legs are larger, your muscle cells have grown and they're contributing to your growing strength.
		  Don't you find you're sweating more? This is another adaptation to exercise. Sweating is a good thing because it helps your body keep its internal temperature in a safe range and avoid the danger of overheating during workouts.
		  Your blood volume has also increased, lessening the prospect of dehydration, another danger to guard against.

WEEK 8

Aerobic Exercise Track	Strength-Building Track	Comments
Drop back to 4 days of aerobics this week, 45 minutes each day. Try something new: Decrease the intensity to 14 during one aerobic workout. On another day, try increasing the intensity to 17. Can you do it?	Do your usual strength-training regimen, but for only 4 days this week.	You may be feeling tired and slightly overwhelmed by your steady exertion of the last 7 weeks. That's why we're factoring in one or two extra days of rest here. We'd still like to see you moving about and working those muscles, however. Use these so-called days of rest to go on a hike in the countryside with friends or a walking tour of some historical landmark. What we don't want you to do is to take us literally—and sleep or stay in bed on your days of rest.

WEEK 9

Aerobic Exercise Track	Strength-Building Track	Comments
On the schedule this week are 5 or 6 days of aerobic exercise at an intensity of 16. For 2 days, you may decrease the duration to 30 minutes, but for the remainder of the sessions, stick with 45 minutes. Option: Begin the week by retaking the Quarter-Mile Walking Test.	This week we move you up to 6 days of strength training. Do the same routines as before, just remember to alternate them.	As time goes on, are you finding you feel more energized? Exercise may make you feel tired immediately after you stop. But its overall effect is just the opposite. It gives people more energy for carrying out the duties of their daily lives. By now your maximal aerobic capacity has improved by almost 10%, which is largely responsible for your increased energy level. Your strength is also showing dramatic improvement due to increases in muscle size and use. A reminder: The cool-down period after you stop exertion is critical. Never skip over it, no matter how much of a hurry you may be in. Slow, languid s-t-r-e-t-c-h-i-n-g exercises, by the way, are the key to reducing muscular aches and pains from other forms of exercise.

149

WEEK 10

Aerobic Exercise Track	Strength-Building Track	Comments
Your assignment: Four or 5 days of aerobic exercise—45 minutes each workout—at an intensity of 15 or 16.	Repeat Chapter 3's tests and reevaluate your strength on 2 consecutive days at the beginning of this week. Then, rest for 1 day. For the remaining 4 days, alternate your upper- and lower-body exercise at 80% of your maximal lifting capacity, which should have changed as a result of your test results.	What about that vital glucose tolerance Biomarker? For sure, your body's ability to utilize blood sugar is up by now. It's up a lot more if you've also managed to cut down on the fat in your diet and reduce your body fat stores. Even if you are losing body fat, your weight may not be going down as much as you expect. This is because you're simultaneously *building muscle*, which weighs more than fat. Don't worry about your weight. What you're doing is extremely beneficial healthwise. While you're shedding fat and building lean-body mass, you're also maintaining your metabolic rate (BMR), which, in turn, assures that you'll burn off body fat more efficiently.

WEEK 11

Aerobic Exercise Track	Strength-Building Track	Comments
This week, if at all possible, we want you to wear that weighted belt we mentioned back in Week 6. Do your aerobic workout for 45 minutes on 5 days at an intensity of 16—and wear that weighted belt throughout. 　Option: Begin the week by retaking the Quarter-Mile Walking Test.	Do 5 or 6 days' worth of the usual strength exercises at 80% of your maximum lifting capacity.	If you're feeling overwhelmed and are having real trouble keeping up with your partner, you are likely going at a pace that's too intense. The Borg Scale is subjective. *You* rate how you feel. Slow down if it's too much. If your partner is going too fast for you, have him or her carry more weight. It will slow your partner down without decreasing the intensity of his/her workout.

WEEK 12

Aerobic Exercise Track	Strength-Building Track	Comments
This week undertake 5 days of aerobic exercise—45 minutes each session—at an intensity of 16. Wear that weighted belt. Make sure that your 2 days off from aerobics aren't consecutive. A good schedule might be Monday, Wednesday, Thursday, Friday, and Sunday, for example.	On Day 1 of this week, retest your upper- and lower-body strength (see Chapter 3). Rest on Day 2, then resume your alternating strength-building routines for the next 4 days. Aim for 80% of your maximal lifting capacity as you do 3 sets of 8 repetitions for each muscle group. Make Day 7 your day off. Your strength exercises should include knee extension, flexion, push-ups, chest exercise, and arm curls.	You are expending between 1,500 and 2,000 calories per week. If weight loss is one of your goals, it should be relatively easy on our program. You'll be surprised at the pounds you can shed over time by following the program, cutting down on fat in your diet, and eliminating your usual between-meal snacks—which were probably high in fat.

WEEK 13

Aerobic Exercise Track	Strength-Building Track	Comments

Another 5 days of aerobic exercise, 45 minutes per day. Change the intensity from one day to the next—14, 15, 16, 16, and 14, for instance.
   Option: Begin the week by retaking the Quarter-Mile Walking Test.
	At 80% of your maximal lifting capacity, do 5 days of strength building. Devote three workouts to lower-body exercises and two to your upper body, alternating them as usual.	If you've gotten this far, you've got a lot of self-discipline and we commend you.
		   If you've been faithful to the program, by now your muscle strength has more than doubled, along with noticeable increases in the size of your muscles.
		   Your training has also resulted in some internal changes for the better that you can't as readily detect. Any loss of bone mineral has slowed or stopped. Your HDL-cholesterol levels are on the rise. And your glucose tolerance has improved, thus decreasing your risk of ever developing diabetes.
		   Remember, rest is a vital part of your weekly schedule. Never shirk it when we offer it to you.

WEEK 14

Strength-Building Track	Comments
Open the week by retesting both your upper- and lower-body strength (see Chapter 3). Rest for one day, then it's back to your alternating routines. Rest on Day 7.	If you want to be both safe and efficient about exercise, we suggest you follow your aerobics sessions with strength training instead of doing them at two separate times each day. Why? For one thing, you only have to warm up once. By the time you get to the strength-building component, your tendons, ligaments, and muscles are less susceptible to injury. In short, it lowers your chances of an exercise-induced injury.

Aerobic Exercise Track
This week's target should be 6 days of walking or biking. Sessions should last for 45, even 50 minutes, not including the time you devote to warm up and cool down.

Something new: We want you to try differing intensity levels during each session. For example, you might start out at an intensity of 13 for 10 minutes, move up to 14 for 10 minutes, shoot up even further to 16 for 20 minutes, then drop back to 13 for the remaining 5 to 10 minutes. |

WEEK 15

Strength-Building Track	Comments
Do 5 days of training, using your upper and lower body on alternate days. For each routine, do 3 sets of 8 repetitions at 80% of your maximal capacity.	By now, we hope you like the weighted belt so much that you wear it during each aerobics session. It's especially effective for those of you who have not graduated from walking to jogging or running. Our Biomarkers Program is designed to get you into shape with as few aches and pains as possible. By carrying weight around your waist, on your back, or in your hands, you can increase the intensity of your exercise, and burn more calories, without having to run.

When you repeat the test for aerobic fitness, you'll notice two things: (1) it's now easier for you to maintain a fast pace; and (2) your heart rate does not climb as much as it did initially. This is because your muscles now utilize the oxygen in the blood much more efficiently. As a consequence, your heart doesn't have to work as hard for you to do the same amount of work. |

Aerobic Exercise Track
Exercise for 5 days—45 or 50 minutes each session—at an average intensity of 15. Alternate the intensity either during the exercise or on separate days.

Option: Begin the week by retaking the Quarter-Mile Walking Test. |

WEEK 16

Aerobic Exercise Track	Strength-Building Track	Comments

It's another week of aerobic exercise, 5 days' worth. Do the aerobic exercise of your choice for 45 to 50 minutes at an average intensity of 15. That's *average*, mind you. Continue to experiment and alternate intensities in the various ways that we've taught you.	Business as usual: Five days of training, targeting your upper and lower body on alternate days. Do 3 sets of 8 repetitions for each exercise at 80% of your maximal lifting capacity.
	
	    At the end of this week, retake Chapter 3's strength tests one final time.
		Congratulations! Just by getting to the point where you're reading these instructions, you've accomplished a lot. After the many weeks of adhering to our program in pursuit of your goal to keep your body younger longer, exercise has become a part of your daily life. Exercise is habit-forming, isn't it? The mere fact that you got to Week 16 means you're extremely likely to maintain your conditioning. With the worst far behind you, you can appreciate how difficult it would be to start over. To avoid ever having to start over, go directly to Chapter 7, which offers guidelines on how to maintain your fitness.
		
		    Faithful adherents to the program now have bodies that have made a remarkable series of adaptations, some of which can be seen with the naked eye and felt every day, others of which are more subtle:
		
		■ Your muscle strength has more than doubled. Your muscles are also larger and you've got more muscle tone.
		■ Your daily calorie expenditure has increased.
		■ You've probably reduced your body fat, provided you also paid attention to your diet and reduced fat food calories over the last 16 weeks.
		■ Your aerobic capacity is greater. You don't get winded or fatigued as easily.
		■ The leeching of mineral from your bones has slowed down, even stopped.
		■ Glucose tolerance has improved, lessening the chances you'll ever develop diabetes.
		■ The level of fats circulating in your blood is in better balance.
		■ Your HDL-cholesterol has increased.

instructions left a little to be desired, you've become a shining example to others.

The last 16 weeks of sweat offers incontestable proof that a sedentary person can respond to exercise in remarkable ways—and that advanced age is no barrier. *At any age, you can increase your functional capacity.* That's what it's all about, isn't it—being able to perform those small daily functions of life easily and without thinking about them? This program has given you added strength and aerobic endurance. Perhaps for the first time in a long time, you can now meet all those everyday challenges involving the lifting of mundane household items without giving them a second thought; and you can walk long distances and really enjoy it.

No doubt your nay-saying friends are impressed with your efforts, even if they're conservative with their praise. Urge them to join you in your endeavors from here on in. The more people you include in your circle of daily exercisers, the less likely it is that you and your companions will ever slip back into your former sedentary habits. To be sure, avoiding recidivism is very, very important, for . . . *your exercise and a nutritious daily menu must continue for the rest of your life.*

No, ours is not a one-shot program like the many weight-loss and instant exercise programs on the market. Our Biomarkers Program is for a lifetime because research shows that conditioned people who want to maintain their fitness must keep exercising. Ironically, when conditioned people stop training, they lose what they've gained at a much faster rate than they gained it—and much more precipitously than those who weren't very fit to begin with. This stands to reason. The more you have, the more you stand to lose.

For recommendations about how to maintain your current level of fitness, turn next to Chapter 7.

CHAPTER
6

BIOACTION PLAN B:
12-WEEK MID-COURSE CORRECTION
FOR THE MODERATELY FIT

———

We use the term *mid-course correction* because it shouldn't take that much to get you back on course —exercising regularly, enjoying it, and reaping the health benefits. We want you to make your exercise program a priority in your life from here on in, while it's still relatively easy and pain-free to do so. Consider the alternative: If you continue with your current lifestyle of irregular participation in sports, your fitness will continue to decline. With each passing year, it will then become more and more difficult to begin an exercise program and set things right. Then, like Bob, the fellow you're about to meet, you'll find yourself guilty of engaging in sporadic exercise without proper warm-up and stretching. Sports injuries, rather than health benefits, will most assuredly be your reward.

Bob's story may sound familiar to you. Like you, he was a

52-year-old, quasi-fit guy when he entered our program several years ago. Like so many people, Bob clung to the mistaken notion that he was in pretty good shape because he exercised every now and then. When he did exercise, he went at it intensely—with an aftermath of physical pain that he could never seem to recover from one hundred percent. Nevertheless, Bob had a vision of himself as an athlete. It was an outdated vision left over from his teenaged and early adult years, and it was leading him astray.

Bob had played on his high school's varsity football team. When he failed to make the college varsity team, he settled for intramural sports—basketball, tennis, some football. But once Bob's school days ended so did most of his sports activities.

Bob fell into the habit of doing what so many of us do—fooling ourselves about how much exercise we're really getting. Bob would give himself credit for exercise because he did some semiphysical chore around the house or at work. For example, he counted as exercise the weekend chore of mowing the lawn. He failed to consider the fact that he sat on a vehicular mower and rode himself around his one-acre lawn.

Bob did participate in a weekly basketball game with the guys, most of whom were half his age. For an hour and a half every Saturday morning, they would go at it like the Boston Celtics, running nonstop up and down the court and engaging in much rough stuff underneath the basket in the absence of any referee to stop it.

When Bob entered our program, he admitted that only sheer grit and determination got him through to the end of each game. After the first 10 minutes, he was gasping for breath—and he suffered the consequences thereafter. By Sunday morning his muscles were really sore, and his right knee and ankle hurt all the time. Bob attributed the pain to advancing age and did his best to forget it and keep on playing with the boys.

Bob's so-called exercise regimen gave us much cause for alarm. Bob was doing everything wrong and was risking serious injury in the process. In his case, our first task was educational.

One bout of intense exercise per week will not increase your fitness. Rather, it increases your chances of hurting yourself—of pulling a ligament or tendon or damaging a joint.

In short, Bob was doing himself more harm than good. What Bob needed was *regular, steady, moderate exercise*. Regular exercise strengthens not only the muscles, but all of the ligaments and tendons that support the joints and attach muscles to bones.

Stronger muscles, ligaments, and tendons are less likely to become injured during any kind of exercise, be it light or intense.

Bob finally accepted that his goal with exercise was simply to strengthen his muscle fibers, not to tax them to the limit on occasion. But it took an awful lot of convincing and six weeks on our program before he began to see the light.

Bob's first reaction to our 12-week BioAction Plan—the one you're about to undergo—was disappointment. He expected more challenge. For the first two weeks he kept asking us how such low-key effort could build anyone's fitness. Nor was he enamored with stretching, which he'd never done before. Stretching took too long, and it wasn't the kind of workout he wanted. Bob wanted things fast and furious, like his weekly basketball game.

Since Bob had such a problem accepting our logic initially, we urged him to trust us. Take what we said on blind faith if necessary, but by all means persevere. We explained that our exercise programs always start slowly. We like to ease people into exercise, not bludgeon them with it. We also assured Bob that, if challenge was what he was after, he'd get it just a little further down the road. About midway through the program, he should start to realize some noticeable gains.

Bob did. At about the five-week point, he found that the intensity and duration of the aerobic and strength-building exercise we'd assigned him started to meet his expectations. He also discovered that his weekly basketball game started to be more of a joy and less of an endurance test. After 12 weeks on our program, Bob found that 90 minutes of running all over the court no longer made him huff and puff at all. While he still wasn't very good at getting the ball through the hoop, he was playing a better game than he had in years. All thanks to regular, moderate exercise off the court every week.

Something else miraculous happened to Bob as a result of our BioAction Plan. His muscle and joint aches and pains disappeared. All those weeks of warming up, stretching, and cooling down properly had had a wonderfully beneficial effect on Bob's musculature. He felt almost nimble again.

AVOIDING INJURY WHILE SHAPING UP

If you're reading this chapter, we know you're moderately fit, but not because you do all that much structured exercise. A weekly golf outing or tennis match, with an occasional evening stroll with your spouse, may be all you can claim as formal exercise. True, you feel pretty good and your doctor says you're healthy. But deep down inside you know that what little exercise you get is not nearly enough.

Instead of ambling alongside that exercise track and flirting with the idea of using it, we want you to hop onto the middle of it and get started down a new pathway to better health and renewed vigor! But we want you to do it in a safe way. Embarking on a new exercise regimen does not have to cause you injury or even much soreness. We are not adherents of the "no-pain, no-gain" school of exercise conditioning. Engaging in exercise without sufficient warm-up, stretching, and cool-down is what leads to damage and soreness. That's why we explained the importance of warm-up, stretching, and cool-down in Chapter 4, and we're stressing it again here. Their value cannot be underestimated.

Ours is a 12-week program. That gives you plenty of time to work up to greater intensities and longer duration. In the beginning of our program, you're going to feel as though your capabilities are being underutilized. You can do more and for longer. Sure you can. But it's a 12-week program, and if you try to transform it into a 2-week program, there's a high probability that you'll injure yourself, or you'll come to hate the program because it's not enjoyable and you'll drop out.

It's easy enough to find excuses not to exercise. "It's raining outside. . . ." "My back hurts today, so why aggravate an already painful situation? . . ." "I'd really prefer a good movie to this routine exertion."

No dice.

Stick to the program and follow our instructions to the letter. For those of you, like Bob, who are chomping at the bit, eager to do more, we have a suggestion. If you must, you can gradually increase the length of time beyond our duration instructions. But *never, ever exceed our intensity directions.*

WEEK 1

	Aerobic Exercise Track — Frequency & Duration	Borg Scale Intensity *	Strength-Building Track	Comments
Day 1:	Walk, cycle, or swim, because these are aerobic exercises that involve major muscle groups. Work out for 30 minutes.	Your intensity target should be around 14. *(Fill in, please.)*		While we'll be giving instructions for walking for the first 2 weeks of the program, you can substitute cycling, swimming, or using an exercise machine. Our instructions about frequency, intensity, and duration are valid for all of them.
Day 2:	Walk for 30 consecutive minutes again today. No stopping during that time to rest.	Your target: 14. *(Fill in, please.)*		
Day 3:	Walk for 30 consecutive minutes.	Exercise at 14 for the first 15 minutes and at 15 or 16 for the remainder. *(Fill in, please.)*	There is no muscular conditioning during the first week of BioAction Plan B.	Remember, at the end of each workout, continue walking at a slower pace—11 on the Borg Scale—for 5 or 10 minutes longer, then stop and stretch (see illustrations in Chapter 4). Besides the half hour of formal aerobics, make an effort to walk more as part of your daily routine.
Day 4:	Add 5 more minutes to your walking session—for a total of 35 minutes, not counting warm-up and cool-down.	Exercise at 14 for the first 15 minutes and at 15 or 16 for the final 20 minutes. *(Fill in, please.)*		This exercise will burn up about 200 calories. That's good. On a weekly basis, this means you're using up more than 1,000 calories. If your body-composition self-assessment indicated the need for body-fat loss, this should come as cheerful news.

* We explained the concept of intensity and reproduced the Borg Scale in Chapter 4.

WEEK 1 (continued)

Day	Aerobic Exercise Track		Strength-Building Track	Comments
	Frequency & Duration	Borg Scale Intensity*		
Day 5:	Rest. There's no scheduled exercise today.			Rest is a crucial part of our program. Making yesterday or tomorrow a rest day instead is not allowed. On the days that you do not perform any formal exercise, try to incorporate more physical activity into your daily routine. For example, park farther away from your destination and walk there briskly. Walk up and down stairs instead of relying on the elevator. At the supermarket, carry a basket instead of pushing a cart.
Day 6:	It's back to 35 minutes of brisk walking.	Goal: 15 _____ *(Fill in, please.)*	There is no muscular conditioning during the first week of BioAction Plan B.	You may find that you and your exercise partner have very different capabilities. One of you may get frustrated trying to slow down to match the other's pace. To even things out, the more robust member of your duo may want to start to wear the weight belt we described in Chapter 4. It will enable that person to slow down and still maintain the same relative intensity as the other partner.
Day 7:	Walk—or ride a bike or swim—for a minimum of 35 energetic minutes.	Goal: 15 _____ *(Fill in, please.)*		By this time it's likely your muscles are sore. It's nothing to worry about. As you continue to exercise, the soreness will gradually disappear. To help relieve the soreness, we recommend warm baths and massage. Do not take aspirin. Your efforts are already having an impact on your Biomarkers. Isn't aerobic exercise a little easier than on that first day? That's because your muscles are adapting and taking up more oxygen. Other positive changes are also occurring. Your muscle carbohydrate stores are greater and your glucose tolerance is improving.

* We explained the concept of intensity and reproduced the Borg Scale in Chapter 4.

160

WEEK 2

	Aerobic Exercise Track		Strength-Building Track		Comments
	Frequency & Duration	Borg Scale Intensity	Frequency & Duration	% of Maximal Lifting Capacity*	
Day 8:	Walk or ride for 40 minutes without rest.	Goal: 15 *(Fill in, please.)*	For 30 minutes, stretch and build your strength. (See Chapter 4 to refresh your memory on the proper warm-up procedure. Stretching and strength exercises are illustrated in the same chapter.) ■ Knee extension: 3 sets of 8–10 repetitions for each leg. ■ Hip & knee extension: 3 sets of 8–10 repetitions. ■ Push-ups: 3 sets of 3; 6 if you can manage it. ■ Now work your biceps: 3 sets of 7 curls.	Your goal throughout the workout: 70–80% *(Fill in, please.)*	Here we go—*your first day of strength training!* As you grow stronger, you'll be lifting progressively more weight. But we caution you to start out on the light side. Men, please don't use this program to act macho. By the same token, we urge our women readers to avoid playing the role of the fragile little lady. Whether you're male or female, 45 or 65, be honest with yourself about your lifting capability. It's very important to use a weight that's right for you—and your strength. If your joints hurt while you're lifting, drop down to a lower weight. Don't forget to warm up, s-t-r-e-t-c-h, and cool down.
Day 9:	REST. If you've been following our instructions faithfully, you deserve it.		Repeat Day 8's strength-training routine.	Goal throughout your workout: 70–80% *(Fill in, please.)*	Get into the habit of associating the intensity of your aerobic workout with a number. Once a certain level of exertion brings a number to mind, you'll find it easier to follow our instructions and you'll get the maximum benefit from the program.
Day 10:	Resume aerobic exercise. Your session should last for 40 minutes.	Goal: 15 *(Fill in, please.)*	Do the same 30-minute strength-building regimen as yesterday and the day before.	Goal throughout your workout: 70–80% *(Fill in, please.)*	By now you should be exercising on a regular schedule—the same time every day. This is key. By adhering to a schedule, your friends and relatives learn your routine and won't interrupt you during that time. An appointed exercise hour every day will probably become a necessity anyway, if you exercise with a partner as we urge you to. *Never schedule your exercise after a heavy meal.* This is unpleasant and just one more way of insuring that you'll give up on your exercise program.

* Maximal lifting capacity, which is the most weight you can lift with one try, is predicted by how many times you lift a weight of X amount. You established your maximal lifting capacity when you took the Strength Test in Chapter 3. To decide how much weight to lift each day, take the goal percentage (see amount listed in Strength-Building Track above) of your maximal lifting capacity. For example, on Day 8 the goal is 70%. If your maximal lifting capacity is 30 pounds, you should lift 21 pounds.

  You'll be retaking the Strength Test every 2 weeks and readjusting the amount of weight that you lift. If you're faithfully adhering to the program, your readjustment should always be upward.

  To continue tracking your progress on the program, you'll need to make photocopies of the empty log book page in Chapter 4 (figure 4-2).

WEEK 2 (continued)

	Aerobic Exercise Track		Strength-Building Track		Comments
	Frequency & Duration	Borg Scale Intensity	Frequency & Duration	% of Maximal Lifting Capacity*	
Day 11:	Forty minutes of aerobics.	Goal: 15 for the entire session. *(Fill in, please.)*	This is a typical lower-body strength-building regimen: ■ Knee extension: 3 sets of 8–10 repetitions for each leg. ■ Hip & knee extension: 3 sets of 8–10 repetitions.	Aim for 70% of your maximal lifting capacity *(Fill in, please.)*	At 70% of your maximal lifting capacity, you should be able to lift the weight of your choosing for 9 or 10 consecutive times. If you can't—4 or 5 times is your limit—then the problem is obvious: it's not weakness on your part, it's simply that the weight is too heavy. By now, you're probably feeling sore as a result of the initiation of strength training. Again, don't worry about it—and don't take aspirin!
Day 12:	Forty minutes of aerobics.	Goal: 15 *(Fill in, please.)*	This is a typical upper-body strength-building regimen: ■ Push-ups: 3 sets of 3; 6 if you can manage it. ■ Now add weights and work your biceps: 3 sets of 7 curls.	Goal: 70% *(Fill in, please.)*	At this aerobic intensity, you should feel comfortable talking to your partner as you exercise. There are two components of your strength-building program: upper- and lower-body training. We want you to alternate between the two. Yesterday, it was your lower body that got the workout. Today, it's your upper body. Stretching becomes even more important as you incorporate strength training into your overall exercise effort.
Day 13:	Ditto yesterday's routine—with one exception. Aim for a greater intensity.	Goal: 16 *(Fill in, please.)*	The focus today is on your lower body: ■ Knee extension: 3 sets of 8–10 repetitions for each leg. ■ Hip & knee extension: 3 sets of 8–10 repetitions.	Goal: 70% *(Fill in, please.)*	By now your strength has improved dramatically. Your brain is "learning" to recruit more of the available muscle cells in your legs and arms. Don't forget to warm up (5 minutes), stretch (5 minutes), and cool down (10 minutes) each time you do aerobic or strength-building exercise.
Day 14:	REST.		REST.		You've completed 2 weeks. The hardest part—overcoming inertia and getting your body moving again—is over. If you were an intermittent sportsman or -woman before starting the program, you probably notice improvement already. You may not be as winded after that tennis or basketball game.

* Maximal lifting capacity, which is the most weight you can lift with one try, is predicted by how many times you lift a weight of X amount. You established your maximal lifting capacity when you took the Strength Test in Chapter 3. To decide how much weight to lift each day, take the goal percentage (see amount listed in Strength-Building Track above) of your maximal lifting capacity. For example, on Day 8 the goal is 70%. If your maximal lifting capacity is 30 pounds, you should lift 21 pounds.

You'll be retaking the Strength Test every 2 weeks and readjusting the amount of weight that you lift. If you're faithfully adhering to the program, your readjustment should always be upward.

To continue tracking your progress on the program, you'll need to make photocopies of the empty log book page in Chapter 4 (figure 4-2).

WEEK 3

Aerobic Exercise Track		Strength-Building Track		Comments
Day 1 & 2: Make aerobics last 40 minutes.	**Day 1 & 2's goal:** 15	This week, alternate upper- and lower-body exercise from one day to the next. You should be doing 3 sets of 8 repetitions of each exercise. You'll be doing two upper-body exercises one day, and two lower-body ones the next.	Goal: 80% of maximal lifting capacity	In Chapter 4, we described both lower- and upper-body strength-training exercises. If you have any questions—or want to try a variation on the upper- or lower-body routine you've been doing—go back and peruse the illustrations and refresh your memory.
Day 3: A 45-minute workout, please.	**Day 3:** 15 or 16		*(Fill in, please.)*	Between your aerobic and strength training, you are now burning more than 1,500 extra calories per week. To lose weight, simply make sure you don't increase the amount of food calories you eat above the level when you started the program. Remember, it takes a calorie deficit of 3,500 to lose a pound of body fat.
Day 4: Lower the duration to 30 minutes but up the intensity to 16 or 17.	**Day 4:** 16–17	Frequency: 5 or 6 days		
Day 5: REST.	**Day 5:** 0	Duration: Each session should take no longer than 30 minutes.		
Day 6 & 7: 40–45 minutes of aerobics.	**Day 6 & 7:** 16			
	(Fill in, please.)			

WEEK 4

Aerobic Exercise Track		Strength-Building Track		Comments
You're going to perform 5 aerobic sessions this week, with 2 days of rest. You choose when to exercise and when to rest. The duration is 45 minutes each session. See the intensity instructions at the right.	Goal: 3 days at an intensity of 16 and 2 days at 15. You choose the days and mix. *(Fill in, please.)*	It's time to find out how you're doing. You're going to retest your strength once every two weeks from now on using Chapter 3's tests. Note the progress you've made. Here's your schedule this week: ■ **Monday:** Remeasure upper-and lower-body strength. ■ **Tuesday:** Work your upper body. ■ **Wednesday:** Lower-body exercises. ■ **Thursday:** No strength training. ■ **Friday:** Upper-body session. ■ **Saturday:** Lower-body training. ■ **Sunday:** Rest.	Goal: 80% of maximal lifting capacity. *(Fill in, please.)*	In Chapter 4, we described a weight belt that you wear while walking. The point of this belt—or a hand weight that you carry—is to enable you to increase the intensity of your aerobic workout without having to increase your speed. Experiment with these weights. Wear the belt, walk at your regular pace, and see what intensity it puts you at on the Borg Scale. As you become better and better trained, we think you'll find you'll want to wear the belt for every exercise session. Your muscles are making marvelous adaptations to exercise, not to mention growing larger. In addition, you've reduced the loss of calcium from your bones; you're burning more calories and, with dietary restraint, shedding body fat; and you're increasing your functional capacity, your ability to do more activities without feeling fatigued.

WEEK 5

Aerobic Exercise Track		Strength-Building Track		Comments
Your aerobic workout this week should, once again, consist of 5 days of exercise. The duration of each session is 45 minutes.	Your training intensity should average 16 *(Fill in, please.)*	Continue to alternate your upper- and lower-body exercise routine from one day to the next. Do 3 sets of 8 repetitions for each exercise. If you have a hard time completing the final set of 8 repetitions, slow everything down. Give yourself a little extra time to recover between sets and repetitions.	Goal: 80% of your maximal lifting capacity *(Fill in, please.)*	You've almost reached the halfway point of your 12-week get-back-in-good-shape program. Last week, you remeasured your strength with Chapter 3's tests. How about testing your aerobic progress this week by revisiting Chapter 3's aerobic test? You should see a dramatic improvement. We generally see a big spurt in improvement in our older subjects soon after the six-week point. Your heart rate should be detectably lower and your walking speed faster.

WEEK 6

Aerobic Exercise Track		Strength-Building Track		Comments
Your goal is to exercise aerobically for 50 minutes, with 5 minutes at both the beginning and end devoted to warm-up and cool-down. That's a total of one hour. Do this 5 or 6 times this week. You choose the days. Rest for 1 or 2 days, performing no exercise at all. Just make sure your rest days—should you decide to take two—aren't consecutive.	Your training intensity should average 16 *(Fill in, please.)*	Retest your strength once again, both to assess your progress and to determine how much weight you should be lifting during the upcoming 2 weeks. Alternate upper- and lower-body routines and continue to work out at 80% of your maximal lifting capacity. Only schedule sessions for 5 days. Take 2-day break from strength training.	Goal: 80% of your maximal lifting capacity *(Fill in, please.)*	Experiment a little with aerobic exercise intensities. For example, start your walk at a Borg Scale 14. After 10 minutes, increase your pace until your intensity reaches 15 or 16 and keep it there for another 10 minutes. Over the next 15 minutes, try for 17, even 18 for a short interval. For the final 5–10 minutes, slow your pace back down to 14. To get a little exercise when you're not formally exercising, wear your weight belt. Why not wear it while you're doing housework or shopping? Among other benefits, it will help you expend more calories, which is what you want to do if weight loss is a goal. Consider the impact all this effort is having on your Biomarkers: The beneficial HDL–cholesterol in your blood is rising, and, if you're following our dietary recommendations, your LDL–cholesterol level is heading south. Your heart no longer has to work as hard during exercise, and your heart rate is lower. Moreover, your total blood volume has increased and you're becoming less susceptible to dehydration.

WEEK 7

Aerobic Exercise Track		Strength-Building Track		Comments
Once again, exercise aerobically (50 minutes a session plus warm-up and cool-down) on 5 or 6 days of this week. Rest the remaining days, performing no exercise at all. Just make sure your rest days—should you decide to take two—aren't consecutive.	Your training intensity should average 16 _(Fill in, please.)_	Train your upper and lower body for 5 days this week, 30 minutes per session. The usual dictates apply.	Goal: 80% of your maximal lifting capacity _(Fill in, please.)_	By now, you have a good notion of what a comfortable aerobic training pace feels like. With that as a baseline, it shouldn't be difficult for you to raise or lower your pace slightly from one day to the next. When you increase your pace—one notch at a time on the Borg Scale, please—your VO_2 will increase at a faster rate than if you maintain a same training pace *ad infinitum*.

WEEK 8

Aerobic Exercise Track		Strength-Building Track		Comments
Our instructions remain the same: Exercise aerobically for 5 or 6 days, one hour's time for a total workout. Rest the remaining days, performing no exercise at all. Be sure your rest days aren't consecutive.	Your training intensity should average 16 _(Fill in, please.)_	There should be 5 days of strength training penned in on your calendar this week. Begin the week by retaking the strength tests in Chapter 3 and readjusting the amount of weight you lift.	Goal: 80% of your maximal lifting capacity _(Fill in, please.)_	By this time, the faithful BioAction Plan followers among you will find that your strength has almost doubled, while your aerobic capacity has improved by almost 10%. Each week, you should be expending between 1,500 and 2,000 calories due to BioAction Plan exercise. For those of you trying to lose body fat, you'll be happy to know that your metabolic rate is humming along at a smart, calorie-burning clip and has not dropped as it would have if you were trying to lose weight through dietary restriction alone. No, a starvation diet is never the answer to the problem of poor body composition.

WEEK 9

Aerobic Exercise Track	Strength-Building Track		Comments
Exercise aerobically for one hour a day for 5 or 6 days. If you're getting bored with the routine, try a new form of aerobic exercise to keep you challenged and motivated. Do not exercise at all for 1 or 2 days, making sure these rest days aren't consecutive.	On Monday, start the week by retesting your strength. If there's significant improvement, adjust the weight you lift upward. Perform the usual alternating strength training for 5 of the 6 remaining days of the week.	Goal: 80% of your maximal lifting capacity *(Fill in, please.)*	By now your muscle strength, size, and overall functional capacity have noticeably increased. Your glucose tolerance has shown a healthy change, your total cholesterol/HDL ratio is lower, the leeching of mineral from your bones has slowed, if not stopped, and your blood pressure is down. All of these changes are welcome. Your higher BMR is making it easier to shed body fat, if that is a goal. We hope you're remembering that our nutritional guidelines are meant to be permanent changes in your approach to food—new dietary *habits*—not just little things you do for a week to see how you like them.
Your training intensity should average 16 *(Fill in, please.)*			

WEEK 10

Aerobic Exercise Track	Strength-Building Track		Comments
Your aerobics agenda remains the same—5 or 6 days of it, then complete rest for 1 or 2 days. The aerobic portion of each training session remains 50 minutes long plus warm-up and cool-down.	Begin the week by retaking the strength tests in Chapter 3 and adjusting the weight you lift. For the remainder of the week, inject a little variety in your routine. Go back over the strength-training exercise modules in Chapter 4 and try some new ones. Just make sure you adhere to our dictate about alternating upper- and lower-body training. The duration remains 30 minutes. Are you finding that your newfound strength makes physical tasks around the house, yard, and office more effortless?	Goal: 80% of your maximal lifting capacity *(Fill in, please.)*	It's critical to replenish body fluids after exercise. Weigh yourself before and after each workout and drink enough of a nonalcoholic beverage to bring your weight back up to where it was. Even if you're not thirsty, drink, drink, drink. Your consistent training over the past weeks has made you more resistant to heat-related injuries. In hot weather, do not be alarmed if you seem to be sweating more than before. It's a good sign that your body is adapting well to the rigors of exercise. Exercise in the cold requires protection for your extremities. Men should wear a jockstrap because frostbite of the penis does happen. After cool-down, your body is wet from sweat and still losing heat. If you don't come in out of the cold without delay, you risk hypothermia (very low body temperature).
Your training intensity should average 16 *(Fill in, please.)*			

WEEK 11

Aerobic Exercise Track		Strength-Building Track		Comments
Do it again, Sam—or Sue. This week, as last, it's 5 or 6 days of aerobics, then complete rest for 1 or 2 nonconsecutive days. Remember, the aerobic portion of each training session is 50 minutes long plus 10 minutes total of warm-up/cool-down.	Your training intensity should average 16 *(Fill in, please.)*	You've got two choices: (1) Continue your strength building as before, or (2) do your entire upper- and lower-body workout in one session. The latter will take close to an hour to complete. Because you can substitute 3 of these longer sessions for the 5 or 6 shorter workouts per week you've been doing, you may prefer it.	Goal: 80% of your maximal lifting capacity *(Fill in, please.)*	You're discovering, no doubt, that virtually all of your physical abilities and skills are improving, from your racquetball game to your gardening labors and heavy household chores. We urge you to wear your weight belt for more activities than just exercise. This will give a big boost to your caloric expenditure and improve your aerobic capacity. If you're still fixated on the readings you get from your bathroom scale, you're probably disappointed. You may be shedding body fat, but at the same time you're adding muscle, which weighs more than fat. Thus, your total weight loss may not be that dramatic. Who cares? Of all the Biomarker changes, increasing muscle is the most important for aging people.

WEEK 12

Aerobic Exercise Track		Strength-Building Track		Comments
You've come to the last week of our formal program to get you in tip-top shape. Once again, exercise aerobically (50 minutes a session plus warm-up and cool-down) on 5 or 6 days. Rest the remaining days, performing no exercise at all.	Your training intensity should average 16 *(Fill in, please.)*	Retake the strength tests in Chapter 3 and adjust the weight you lift accordingly. For the remainder of this week, your strength-building regimen remains the same. Continue to alternate the upper- and lower-body workout, whether it's during 3 long sessions of an hour's duration, or twice as many half-hour sessions. If the hour-long sessions become your practice, never strength train more than 3 days a week. Your body requires sufficient time for recovery.	Goal: 80% of your maximal lifting capacity *(Fill in, please.)*	Do you feel noticeably stronger? No doubt your muscles are now larger, your body fat is lessened, and you can perform with relative ease activities that previously caused you dismay—or you couldn't do at all. This accomplishment calls for some kind of celebration. After all, if you maintain this fitness over the years, your life expectancy will increase and the quality of your physical life will be the envy of your friends. For sure, the people who love you understand what you've done and will thank you for years to come.

CHAPTER 7

BIOACTION MAINTENANCE GUIDELINES

———

You're reading this chapter because you've passed a milestone.

Some of you picked up this book knowing you were in poor condition. You took our self-tests back in Chapter 3 and got verification of your worst suspicions. That—and what you learned about those inevitable Biomarkers back in Part One— spurred you on. You followed our 16-week BioAction Plan for the low-fit. It wasn't easy, but here you are—a graduate and proud of it!

The rest of you picked up this book with a take-it-or-leave-it attitude. You've always paid cursory attention to health matters. Many of you have even been known to take up exercise for periods of time. But, by and large, consistency of physical effort has never been one of your virtues.

No doubt you started reading this book because you figured there might just be something more you could learn. Then, when you didn't do brilliantly on Chapter 3's self-assessment, you got worried enough to sign on to our 12-week program. You, too, ought to feel very proud of yourself—and determined not to backslide after all the effort of the last few months.

Having arrived at this point, all of you want the answer to the same two BIG QUESTIONS. The first . . .

HOW MUCH AEROBIC EXERCISE IS ENOUGH TO MAINTAIN MY IMPROVED STATE OF FITNESS?

No lesser an authority than the American College of Sports Medicine (ACSM) states that your aerobic fitness can be maintained with as little as 30 minutes of exercise three days per week at an intensity of about 15 on the Borg Scale. Be aware, however, that *this is a minimum amount of exercise*. It's actually less aerobic exercise than you were getting in the closing weeks of our two BioAction programs. Should you fall below this minimum ACSM recommendation, your fitness level will definitely decline. That's why our recommendation is a little more stringent:

We want to see you continue working out aerobically 4 days per week for 45 minutes per session at an intensity of 15 or 16. This is particularly important if you've lost weight on our BioAction Plan and want to keep it off. (Who doesn't?) The more you exercise, the more calories you'll burn. It's really a matter of simple arithmetic.

A recent large-scale study, carried out by the Cooper Clinic and Institute for Aerobics Research in Dallas, underscores just how important continuing aerobic workouts are for your long-term health—and how even a little of this type of exercise goes a long way. The study, the largest to date to relate physical fitness to mortality, followed more than 13,000 men and women for an average of eight years.[1]

The study participants, all relatively healthy, were grouped into five aerobic fitness categories, ranging from low to high, based on how they scored on treadmill stress tests. In general, the least fit were people who did no formal exercise. The most fit worked out long and hard, some running as much as 30 to 40 miles a week. The middle groups had active lifestyles that usually included some exercise.

Study findings: Those who lasted the least amount of time on the treadmill—those in the lowest fitness category—had a far greater chance of dying prematurely from a chronic disease than those in all the other groups. When researchers compared groups, the men in the highest fitness category had death rates that were about three and a half times lower than men in the low-fitness group. Among the women, the middle-highest fitness group actually had the lowest death rate. (Their mortality rate was not appreciably different from the highest-fit group.) It was some six times less than that of the lowest-fitness group.

Here's the most exciting study conclusion for those of you who aren't ardent exercisers:

The biggest improvement in the mortality-rate picture comes from moving out of the lowest-fitness, sedentary category into the next category—and it doesn't take much to make that transition. According to Dr. Steven N. Blair, the lead researcher on the study, a 30-to-60-minute daily walk, brisk but not to the point of discomfort, will do it.

Still, these Cooper Clinic and ACSM guidelines represent the least amount of aerobic exercise you should be doing. It's up to you to decide the maximum amount.

HOW MUCH STRENGTH CONDITIONING IS ENOUGH?

Our recommendations, based on our own research at Tufts, are simple[2]:

To maintain the muscle power you've gained, you should continue training three days per week, keeping the weight the same as in the last week of the program and alternating upper- and lower-body workouts.

On the other hand, if you like the new, more toned and muscular contours of your body, you'll want to push ahead and add to your muscle size and strength. To achieve this goal:

Continue to lift greater and greater weight, but gradually, very gradually. At all times, your intensity should remain at about 80 percent of your maximal capacity; you should exercise three days per week.

As you venture out into the world of exercise self-reliance, be aware that it's oh so easy to fall back into old sedentary habits. That's why we feel the best motivator to help you make exercise a part of the fabric of your everyday life is this startling fact:

The effects of exercise are short-lived.

If you stop exercising after completing our Biomarkers Program, within one week your insulin sensitivity will return to pre-training levels and many of the remarkable adaptations your muscles made to exercise will be well on their way to being lost. No, the beneficial effects of exercise can't be stored. Exercise and body movement are intangible experiences that your body needs to undergo daily. You might even say that activity, along with food, are the two fuels your body's engine must have in order to function normally. If you continue to eat but halt activity—stay in bed for a number of consecutive days, for example—you'll find out just how much your body really does require movement to function properly. This is why we say our Biomarkers Program is a program for the rest of your life.

Questions from You, Our BioAction Graduates

We'd like to share with you some of the questions we hear most often from graduates of our program—as well as a few of the initial excuses we've heard from people who eventually entered the program with fine results:

Q: _I went through your program and was following the maintenance guidelines for several months—until a tragic death in my family and all the emotional and financial turmoil because of it threw me off course. I stopped exercising completely for a month. Does that mean I have to go back to square one and start all over?_

A: It means you have to go back to Chapter 3 and retake the self-assessment tests to find out what shape you're in. Your scores on those tests will slot you into one of our BioAction Plans. Yes, you should restart your exercise program gradually. Following a BioAction Plan is the safest way we know to do this. In your case, since you stopped only for a month and were a faithful exerciser before that, you'll probably be destined for BioAction Plan B for the Moderately Fit.

Q: _You stress that two or more people should always exercise together, so my wife and I keep each other company—and bolster each other when one feels like quitting. However, we have a problem with the aerobic portion of the program. I'm 6 feet 1 inches tall and my wife is a petite 5 feet. When we walk together, my stride—even when I try to slow down —puts me way ahead of her. What should we do?_

A: One of the reasons we recommend that you purchase weight belts is to equalize this very situation. If you're much faster than your wife, the extra weight around your middle will slow you down. It will enable you to get a good aerobic workout at your wife's slower pace.

Here's something else to bear in mind: Exercise partners must begin our program at the same time. If one partner starts after the other has been training for four or six weeks, the partners have no commonality of experience. The state of their fitness is too disparate.

Q: *While I know that I should begin an exercise program, my arthritis stops me. It hurts just sitting still, and it keeps me from doing so many things. How can I possibly exercise?*
A: If arthritis is a major problem for you, yes, we agree you should proceed with caution.

The first thing you should do is get a medical exam and have your doctor confirm that arthritis is your problem. Many older people have a condition of the hip or knee called "degenerative joint disease." Should this turn out to be your real ailment, ask your physician what he or she recommends. A referral to a physical therapist may be in order. A walking program, such as the one that's an integral part of our Biomarkers Program, is probably *not* a good idea. It could add to your joint problems.

On the other hand, if arthritis is truly your malady, exercise is certainly not out of the question. It will just be a more protracted process for you. You'll have to warm up for a longer period than most people, but the good news is that the warming effects of your exercise will probably lessen the pain both during the session and for a short "halo" period afterward.

We had a woman with rheumatoid arthritis undertake our walking program with marvelous results. She was in such pain sitting around doing nothing, she finally decided she had nothing to lose by attempting aerobic exercise. She discovered, to her happy amazement, that it didn't hurt while she walked. She found the surcease from pain such a relief that, before long, she was walking up to five hours a day!

While we cannot promise you that all arthritis sufferers will share this woman's experience, there appears to be sufficient evidence that walking has pain-relieving and other beneficial effects to warrant giving it a try.

An aside: Many arthritis sufferers take hot baths or saunas or soak in hot tubs to relieve their pain. If this is your habit, fine. Just don't break the cool-down rule we laid out in Chapter 4: No hot showers, baths, steam rooms, saunas, and so on for at least 45 minutes after you've completed exercise.

Q: *I set aside an hour and a half a day, five days a week, for exercise. I do all portions of your program—stretching, weight training, aerobics—during the same workout. Even though I'm pretty tired at the end, I don't have time to do it any other way. What confuses me still is the order. Is there an optimum sequence?*

A: Yes. Combining all the different forms of exercise into one long session is fine, provided you follow this recommended sequence: warm-up, stretching, aerobics, strength training, and cool-down.

This sequence involves a progressive warming of your muscles. The stretching increases your muscles' flexibility before you embark on the more intense aerobic phase. In turn, aerobic exercise is an excellent way to warm up your muscles even further before strength training. Lifting weight when your muscles are warm greatly reduces the risk of pulling a muscle, tendon, or ligament.

Q: *When my exercise companion and I first assessed our strength using your self-tests, we shared the same low score. Then we started your program and we discovered my friend's muscles respond to strength building much better than mine. Every time he retook the strength tests, he always showed much more improvement than I. He'd immediately boost the amount of weight he was lifting and be even farther ahead of me the next time we retested ourselves. I was trying as hard as he was, but I didn't realize similar results. How do you explain the disparity?*

A: Your experience is not unusual. Everyone responds to weight training differently. The reason for your friend's remarkable improvement in strength is most likely related to his muscle fiber type.

As we explained in Chapter 2 when we discussed the strength Biomarker, our muscles consist of fast- and slow-twitch fibers (which is another name for muscle cells). The amounts of these two types of fibers in your body is a gift from Mom and Dad; it's genetically determined. If you've got a higher proportion of fast-twitch fibers, you'll probably see a faster rate of improvement in your strength during a muscular conditioning program. Why? Because fast-twitch fibers increase in size faster in response to weight lifting.

You shouldn't feel that Mother Nature cheated you, however. People who don't respond as well to strength training usually respond with greater ease to aerobic exercise.

Q: *After a slow initiation into exercise—the first six weeks were murder!—I've finally come to appreciate how much better a workout makes me feel, not to mention its wonderful tension-releasing effect. It's gotten to the point where I don't feel right on days when I don't exercise. Do I really have to take at least one day off each week for rest?*

A: YES! YES! YES! At least one day a week of complete rest from exercise is important. Follow our advice and you'll notice a payoff. The day after your rest, your exercise will feel easier than usual. No, it's not your imagination. This is because the short rest period allows your body time to recover from the trauma of repeated days of exercise stress. Your muscles have time to refuel and repair some of the small damage that occurs during workouts.

Many serious athletes who exercise every day without letup learn to live with continual aches and pains. In some cases they also accept a chronic feeling of fatigue. This is just the opposite of what we want our Biomarkers Program to do for you.

Q: *Could you settle a long-standing controversy between me and my jogging companion? He says speed is of the essence. I say slow and steady wins the race—that the distance covered without stopping is more important. Who's right?*

A: You're both right. The answer really depends upon your reason for exercising.

If you're concerned about burning more calories, the total distance covered is what's most important. Whether you jog slowly and steadily or sprint in covering one mile, you still burn roughly 100 calories. However, you would expend slightly fewer calories per mile if you simply walk.

Speed becomes the issue if you're a competitive athlete and want to increase the pace at which you run. Repeated bouts of high-intensity, sprint-type exercise to exhaustion—called "interval training"—will help you do this.

If you're interested in exercise only as a means to make you healthier and more fit, forget about sprinting or interval training. Sprinting is difficult to do on a regular basis, and there's no evidence that it has any greater long-term health benefit than steady jogging. In fact, unless you're sufficiently well trained, high-intensity exercise only increases your risk of injury.

Q: *I'm a salesman for a drug company, and I have a city territory. For long hours every day, I lug around a 30-pound sample case from one office or pharmacy to another. Mostly I walk, except in bad weather.*

Doesn't this job-induced exercise count for something? Doesn't it mean I need less off-hours exercise?

A: Of course this activity counts. After all, your body doesn't differentiate between movement that's work-related or movement that's recreational. We commend you for walking instead of riding to appointments.

For some occupations that are very physically demanding, on-the-job work may be all the exercise that's needed. If you, for example, carry your 30-pound case and walk for 20 to 30 minutes *without stopping,* twice a day, for a total of an hour every weekday, yes, you should be getting all the aerobic exercise you need. One way to find out is to measure your aerobic capacity with the aerobic self-test in Chapter 3. You should fall into the good-to-excellent category.

Q: *I'm only 55 years old and hardly decrepit. I followed your 16-week program with a friend who is 10 years older. During the aerobic portion of each session, we aimed for the same intensity goal, the one you suggested. What bothered me throughout was that I always seemed to reach the intensity target much sooner than she did. She could do much more exercise before she was exercising "very hard." What's wrong with me?*

A: Nothing is wrong with you. It's just that your friend is better trained than you are. She probably started off with a higher aerobic capacity than you—or she may be doing more exercise during the time when you're not together. She may be doing more errands on foot, exerting herself more with gardening or housework, for example.

We realize it's demoralizing when exercise partners aren't equals. If you and your friend are to continue to work out together, we suggest that your friend wear a weight belt or carry weights in her hands or around her ankles. This will slow her down so that your pace and intensity are equalized.

Q: *I can't do the aerobics and muscular conditioning all in one session. It tires me out too much. I do them separately, which brings up these two questions:*

Which type of exercise should I do during the earlier session of the day?

Should I combine the stretching with the aerobics or with the strength-building portion of the program?

A: It's a good idea to split up your workouts. Whenever you feel too tired to continue any exercise, your first response should be to reduce the total amount of time—the duration—of that exercise.

In answer to your first question, it really doesn't matter which type of exercise you do earlier in the day. When should you stretch? Ideally, before both the aerobic and strengthening portions of the program. However, if you find this too burdensome and you have to make a choice, stretch before the strength-building session to reduce the risk of injury.

Q: *I travel a lot for my job. Not all the hotels I stay in have health clubs, making a workout very difficult. What's your solution? Could I, perhaps, double up on the amount of exercise I do on weekends to make up for the days I miss during the work week?*
A: Doubling up on your weekend exercise is not a good idea. Our program emphasizes regularity of exercise. If you exercise for two hours one day, followed by no exercise for the next four days, you'll find it exceedingly difficult to maintain your fitness. You'll also be greatly increasing your chances of sustaining an injury.

We recommend walking as the aerobic exercise of choice because of its universality. It can be done almost anywhere, anytime, as Ralph Waldo Emerson, the nineteenth-century American philosopher, once pointed out: "The qualifications [of a good walk] are endurance, plain clothes, old shoes, an eye for Nature, good humor, vast curiosity, good speech and good silence."

Walking is a particularly delightful way to see a new city when you're traveling. In contrast, if you rely on exercise equipment as the means to stay fit, we concede you've got a problem when you and the equipment are separated.

What you may want to do is develop a repertoire of aerobic exercises you like. That way, you can switch as the situation demands. Most hotels have pools, so if you don't want to walk or jog in a strange city, you can take a swim instead.

Remember, the minimum amount of aerobic exercise you should be performing is 30-minute sessions, three days a week. If you miss a day or two a week because you're on the road or working too hard, don't feel too guilty (just a little guilty).

In contrast, maintaining your muscular conditioning while you're on the road should pose no problem whatsoever. As we pointed out in Chapter 4, special kinds of rubber bands, called "Dyna-Bands," are the answer to every traveling exerciser's prayers. These stretchy latex "weights" are inexpensive and about as portable as you can get.

Q: *I don't own an exercise bike or any of the other indoor exercise equipment on the market. Without investing in any of these contraptions, would you kindly tell me how to discharge my daily aerobics obligation in inclement weather?*
A: Many people use weather as an excuse not to exercise. During our walking studies with women subjects, we provided a gym for them to exercise in during inhospitable weather. In more than two years of our program, our women have never used the indoor facility—even during two cold Boston winters. They dressed appropriately and preferred to brave the cold because they felt walking around and around a small gym would be too boring.

On cold winter days, they reported that their exercise, after the first 5 or 10 minutes, generated enough heat to keep them comfortable. It's true, they didn't venture forth during blizzards and driving rains—but how many days like this are there every year? During the summer months, they exercised in the relative cool of the early morning.

Q: *I finished your 12-week program feeling good, but the moment I didn't have your chart to look at for instructions each week, I found it harder to stay motivated. Are there any motivational tricks for the long haul that you can share with me?*
A: This is the chief reason we recommend that you work out with a friend or group of friends. When you exercise alone, it's easy to find excuses. When you have people waiting for you, you can no longer play this game with yourself. Your exercise companion(s) call your bluff.

In the next chapter we outline a medley of popular motivational tricks to keep you exercising. There, we describe what's worked for the thousands of older people we've counseled over the years. Broadly, we've found the following:

Super achievement–oriented people often spur themselves on by setting goals. For them, exercise becomes a means to an important end. Invariably, these are people who love challenges and savor victories. When they finally achieve their goal, they relish the feeling of success. They have another notch in their accomplishment belt.

Other people respond better to rewards. For example, such

people might keep track of the distance they walk and, after every 500 miles, treat themselves to a night out at their favorite restaurant or a new possession they crave. We've even known people who set up a savings account and place a dollar in it for every mile they walk or every strength-building session they undergo. That's where the money eventually comes from for that special reward, whatever it may be.

Q: *When I go on vacation, can I take a vacation from formal exercise too? After all, I'll be doing a lot of walking to see the sites.*
A: Many of our program's graduates plan physically active if not downright demanding vacations. They choose to go biking in Ireland or hiking in the White Mountains of New Hampshire rather than taking a slothful cruise or basking in the Caribbean sun. The last thing you should do is lie for two straight weeks on a beach, where you'll lose most of the fitness you've gained. By all means, plan a lively vacation that has you and your companions exercising—biking or walking—to see the sites every day.

Q: *Can I splurge on food when I'm traveling in some exotic part of call, especially if I promise to return to my more prudent dietary ways after I return home?*
A: Of course you can. There is nothing more enjoyable than tasting the local cuisine when you're in faraway places. Be forewarned, though, that many peripatetic people who eat in restaurants a lot have a very difficult time maintaining their diet. Don't enter their ranks. Don't let travel become your excuse for not following a good diet plan. One day of splurging every couple of weeks is quite different from two or three days of it per week.

CHAPTER
8

OVERCOMING MENTAL OBSTACLES ABOUT EXERCISE

By this point in the book, you probably think you need no more convincing. You're determined to stay out of that Disability Zone as long as possible—even bypass it entirely—by turning back, or at least slowing down, the tick-tick-tick of your own biological clock. You've digested the evidence about those inevitable—though largely *controllable*—Biomarkers of vitality, and a little voice in your mind is vowing, "I'm not wasting any more of my precious time on this earth. From here on in, I'm going to do my level best to follow the exercise program and dietary recommendations in this book. I see the error of my ways and I will do everything in my power to right the wrongs of the past."

Would that it were that easy.

Certainly, a thorough understanding of why you should become more active and a resolve to do it are crucial first steps. We think we've helped you see the need for exercise in the opening chapters of this book. But, as many a sage has pointed out, the road to hell is paved with good intentions. You're about to discover that making the transition from head talk to muscular action is a major challenge. If the experiences of our study volunteers are at all representative, we predict you'll experience a flood of alien emotions during the transitional period when you're trying to develop a daily exercise habit.

We often ask participants in our exercise studies to fill out anonymous questionnaires at the end of their first few weeks of increased activity. Basically, we want to know how they feel about the changes they're undergoing. Besides the expected complaints about sore muscles and the unpleasant sensation of being winded, we're constantly amazed at the outpouring of emotion that exercise seems to engender. Here are some typical comments:

Embarrassment: "When I lift weights, I feel as if the cameras of the universe are recording my every awkward move."

Humiliation: "I've never smoked in my life. So why does a little aerobic exertion leave me coughing and gasping after just 20 minutes? How did I ever get so out of shape?"

Self-deprecation: "Walking very fast all dressed up in my new exercise duds makes me feel foolish, even if I know it's good for me. I'm 72 years old, after all."

Confusion: "I feel fat, wheezy, sweaty, ashamed, hopeful, discouraged, angry, silly, like this may be the best thing that ever happened to me. None of this makes sense. I feel like a bundle of contradictions."

Self-doubt: "My friends think I'm crazy. Don't I have any gumption or willpower? Why do I need a regimented exercise program to make me move my carcass?

Vanity: "I've always hated exercise. I still hate exercise. But if this helps me lose weight, that's all I care about. Nothing else has worked very well, so why not give this a try? I've got nothing to lose but some fat, if I'm lucky."

Misplaced identification: "The walking is fine. But to me the muscular conditioning has negative connotations. I keep picturing macho, muscle-bound men like Arnold

Schwarzenegger. I don't want to look like that or grunt and strain the way they do."

Clearly, any battle older people are having with their bodies in the beginning is nothing compared to the war going on in their minds. Some find themselves wrestling with many unresolved psychological issues involving health and exercise. This is especially true of today's over-65 generation who were born in the first quarter of this century when medical practice, not to mention social customs and mores, were radically different. They see the manifestations of the fitness revolution—younger adults jogging along the side of a road, filing into local health clubs, bicycling to work in their business suits in lieu of a bus ride—but it's not something they necessarily identify with. They know what's happening and why, but deep down inside many feel that exercise is for the middle-aged man who wants to forestall a heart attack or the woman obsessed with her figure and attracting a man. They think, "I'm beyond all that."

UPDATING YOUR ATTITUDE ABOUT EXERCISE

Maybe you didn't grow up with the idea that exercise should last a lifetime. In your day, sports were meant to last until you graduated from school, possibly until middle age—but never, ever beyond that.

Whether you realize it or not, such thinking may well be the reason you've become so attached to that comfy, old easy chair and so resistant to the idea of more movement and exercise in your life. After reading the opening chapters of this book, we assume you now realize why such outmoded thinking—especially in light of our research to the contrary—is hazardous to your future health!

We've devised a little visualization to help you unearth what you really feel about exercise in all its myriad forms. Do this mental exercise when you're alone and in a contemplative mood. In ruminating upon the scenes we describe, let your mind form pictures. Don't censor what floats across your mental movie screen. Watch the images and let your imagination unfold. See and hear. Don't think and reason.

• You enter a room filled with barbells and elaborate Nautilus machines. At each exercise station there's an elderly man or woman pumping away, miraculously defying the frailty of their appearance. You amble around the room, exchanging banter with each of them as they stop for a rest. Listen to the conversations.

• You're on a California surfing beach in the summertime. For once, it's almost deserted since a storm is threatening and the sea is rough. A young, blond-haired surfer, around 20 years old, is the only one skimming the waves. He seems oblivious of the frothy turmoil—or is he hell-bent on proving that mankind, through sheer determination and willpower, can conquer the raw force of nature? As you watch and contemplate the young man's motivation, you see him take a violent spill and disappear under the water. Before you can decide what to do, a gray-haired, compact elderly man, whom you hadn't even noticed, hurries into the ocean. How can this older man save anybody? To your surprise, once he's in the water, his movements become dexterous and fluid. He swims with the strokes of a trained swimmer. You finish the story.

• You're waiting at a bus stop in front of an urban nursing home. It's one of those early spring days when the flowers are starting to bud, the temperature is mild, and the world hates to be indoors—including the feeble denizens of St. Hilda's Rest Home. A young woman, dressed like a nurse except for her track shoes, bursts through the front doors. Like the Pied Piper of Hamelin, she's followed by a crowd of chattering senior citizens, out for their daily "power walk." Once everyone is outside, the whistle-blowing nurse begins to bark out orders like a high school football coach. How do the old folks respond? How do you and other passersby respond?

There's no right or wrong way to conclude any of these scenarios, as we're sure you realize. Our purpose with the above was merely to get you to confront some of your innermost feelings and become aware of any resistance you may still harbor about the dynamic duo of exercise and older people.

MORE RESISTANCE IN THE BEGINNING, GREATER COMMITMENT IN THE END

It's true that you, as an older person, may be slower to respond to exhortations about exercise than the typical young adult.

But, if our experience is any guide, when you finally do decide to try it, you'll be far more committed.[1]

It may take less convincing to get younger people, with their animal spirits and high energy levels, into exercise clothes, but their resolve peters out faster, too. They jump right in and start overdoing it immediately. Then, after a few short weeks, they're worn out, bored, and itching to get involved in the next fad on the horizon. They rush ahead without giving what they're doing a whole lot of thought; and, just as abruptly, they become dropouts.

In contrast, older people tend to circle around the idea of exercise for a while before they commit themselves. But when they do, it's a more wholehearted commitment. Older people have a greater reservoir of reference points, and they understand the value of forethought. In contrast to younger people, with their cavalier attitude about the passage of time, older people realize that a lifetime is finite and that the time left is precious. They cannot afford to waste effort on frivolous endeavors.

Even if it sometimes takes more patience to deal with an older person's initial rigidity about sports and exercise, we prefer working with a mature age group, both for the reasons we just cited and because, from a scientific point of view, it's more rewarding. As we've said throughout this book, no group of human beings stands to benefit more from vigorous exertion.

BREAKING THE HABIT OF IMMOBILITY VIA THE BIOMARKERS PROGRAM

Despite what we just said about older people and commitment, expect to have a hard time during the first few weeks on the Biomarkers Program. No matter what your age when you start an exercise program, the beginning is always the hardest part. We apologize for the cliché, but there's no better way to say it. You're trying to change a habit, after all—*the habit of not exercising*. Like any other habit, it will take a concerted effort and a lot of self-awareness on your part in order to break it.

There's a sound psychological reason why our two BioAction Plans—for the low fit and moderately fit—are 12 and 16 weeks long, respectively. Behavioral psychologists tell us that it takes 21 days to establish a pattern and 100 days to make it automatic. One hundred days is about 14 weeks. It's been our personal experience that people who get beyond an initial three-month threshold period in an exercise program or diet change stand an extremely good chance of continuing the new pattern thereafter.

What you must realize is that _our Biomarkers Program is not a quick fix_. You can't follow our program to the letter for three or four months, pat yourself on the back for a mission accomplished, and then forget it all and go back to the way things were. Frankly, if that's what you've got in mind, we suggest you put this book down right now and pick up a romance novel or mystery story. Either one will probably do you as much good.

We'll be honest with you: Four months of exercise via our program isn't going to do all that much to reverse the onrushing tide of those inevitable Biomarkers. As we've said several times before, _the effects of exercise are short-lived_. If you stop exercising after completing our Biomarkers Program, within one week your insulin sensitivity will return to pretraining levels and many of the remarkable adaptations your muscles have made to exercise will begin to dissipate.

The only thing that will have the positive impact you desire is _a total behavioral change_. And that new style of life must be just what we said—_daily_. It must be the way you conduct yourself every single day for the rest of your life!

MOTIVATORS TO GET YOU OVER THE FIRST HURDLES

As we've said, the early days on any exercise program are critical. It's the time when the drawbacks still loom larger than the rewards and excuses flood through your mind at the mere thought of putting on your walking shoes or picking up a weight.

During the opening weeks on the program, it's essential that you pick up this book daily. In our BioAction Plans, we offer detailed instructions about how to proceed through the physical aspects of the program. Read them carefully to make sure you're doing everything right. Then reread the motivational words that follow to boost your incentive. Commune with us, view us as helpful companions on your exercise odyssey. We want you to hear the echo of our words in your head as you experience the triumphs and travails of your initial exercise bouts.

The following motivators are intended to keep you up and at 'em. They've worked for the older people we've counseled, so we're confident they'll strike a resonant chord with you, too.

• **Be clear about who is in control of your life.** The circular reasoning that comes from a passive attitude toward life is a serious roadblock to success in any endeavor, including our Biomarkers Program. Until you start erasing the excuses and "I can'ts" from your speech and accept the fact that *your health is your responsibility,* those inevitable Biomarkers will proceed apace and you'll keep wondering why your more active, less self-pitying friends feel better than you do.

Social psychologists sometimes make the distinction between "inner-oriented" and "other-oriented" people. People who are inner-oriented are self-reliant and accept responsibility for what happens to them in life, sometimes to a fault. In contrast, other-oriented people see the source of control existing outside themselves. When things go wrong or life doesn't work out as planned, other-oriented people are more likely to lash out against those around them, rail about circumstances or brood about bad luck. Inner-oriented people will turn the blame inward and scold themselves unmercifully.

To be sure, neither extreme presents the ideal picture of mental health. We mention it here because these two kinds of people come around to the idea of a lifestyle change from different vantage points. Once convinced of the efficacy of exercise and a better diet, inner-oriented people are likely to act as though they invented the idea. As such, they're more likely to stay motivated since it was their decision to do it in the first place. They take full responsibility for their decisions and actions, be they right or wrong. Of their own free will, they are *choosing* to have a healthier lifestyle. Their problem comes later when they experience their first setback or fail in their resolve; we'll discuss this in a moment.

Other-oriented people can also be convinced to exercise and reorder their eating, but they're more likely to say they're doing it because their doctor, or a book, or their friends talked them into it. These are people who have more of a problem staying motivated because there's no inner fire involved in their decision. They're just doing it because it seems sensible and it pleases someone else. They're doing it out of a sense of duty.

If you see a lot of yourself in our description of the other-oriented person, you may have a history of waiting around for lightning to strike. You persist in thinking that some nostrum or pill or new medical technique will eventually make you look and feel younger. Like the dilettante, you've probably dabbled with other fitness programs and will likely jump on the next anti-aging bandwagon that comes along.

If your pattern is that of the experimenter rather than the

committer, we suggest you take serious stock of your situation. Keep the words of Ern Baxter in mind. He wrote a book called *I Almost Died* about his recovery from a midlife heart attack that frightened him more than he ever thought possible. He writes, "If your lifestyle does not control your body, eventually your body will control your lifestyle. The choice is yours."[2]

• **Know the *real reason* you're embarking on the program.** There are hundreds of reasons why someone might embark on an exercise/diet program. Your reason won't be the same as the next person's—that is, if you're being truly honest with yourself.

It's our experience that people who cite noble reasons for changing their habits—"Because it's good for me. . . ." "I'll live longer. . . ." "It will postpone my entry into the Disability Zone"—are people who have a hard time staying motivated. Human beings, as thinking animals, need concrete reasons for doing things—and they need tangible proof of progress. The above reasons are too sweeping, and the proof that exercise is bearing fruit is too far in the future.

To be of any value, you must state your reason in terms of the here and now. Your reason must be something that has personal significance to you and that you can get feedback about within a matter of months, not years or decades.

What are some real-life, easily measurable reasons? Here are a few that older people, once we gained their trust, have shared with us:

• "Now that I'm retired, I have a lot of pent-up aggression that I didn't feel when I was working as a plumber. But after I went from that physical job to sitting around all day, suddenly I'm angry all the time. I figure exercise may help me with this problem." (66-year-old man)

• "I need companionship and to get out of the rut I'm in since my husband left me. There are a lot of people my age involved in your program. I think coming to your center and participating in your research studies will put a spark back in my life." (52-year-old medical records administrator)

• "When I was younger, I used to exercise regularly. But then I got married, had kids, and I got busy doing other things. Lately, I have this uncomfortable feeling that circumstances and life are controlling me, that I'm no longer master of my own fate. I want to prove to myself that I still have the willpower and self-discipline to follow a formal exercise program, especially now that I'm older and maybe it won't be so easy." (55-year-old male schoolteacher)

• "Five years ago, my doctor told me to stop smoking, lose weight, and reduce the stress in my life, even if it meant finding another occupation. I have several risk factors for heart disease and early symptoms. I didn't do all that he advised, and, sure enough, I had a heart attack. I barely made it. I learned something from the experience, though: I want to live. And if exercise and better eating habits will prolong my chances of being around a while longer, that's a small enough price to pay." (61-year-old former air flight controller)

• "I broke my leg six months ago. It was in a cast for what seemed an eternity, and when it came off it was so weak I couldn't stand on it. The doctors gave me exercises and I did them faithfully every day. It wasn't long before my fractured leg was in better shape than my uninjured leg. I asked my doctor about it and he assured me my whole body could regain the muscle power of that leg if I was willing to do regular strength-building exercises. So I'm signing on to your program. The rejuvenation of that leg made me a believer. I want the full makeover." (69-year-old housewife)

You may find it takes a little time to cut through the self-deception and figure out your true purpose for going on the program. But it's important that you go through this process of stripping away the falsity and getting down to true intent. Your real reason may be anything but physical or health-oriented. It could be emotional or purely social. That's fine. Just know what it is, and once you do . . .

• **Sign a contract with yourself.** State your reason *in writing* for going on the program. Put it in terms of a tangible goal that can be at least partially realized by the time you complete our formal 12- or 16-week BioAction Plan.

An example might be: "I am exercising to reduce stress, feel more relaxed, and cut down on the alcohol I consume each evening."

The idea of "signing up" is the brainchild of Leonard Wankel, Ph.D., the graduate program director of the Department of Recreation and Leisure Studies at the University of Alberta in

Canada. He used this technique in a six-month cardiac rehabilitation program with good results. The participants fell into three groups. The first consisted of those who signed a statement outlining why, when, and how often they would exercise. Sixty-five percent of this group stuck with it. The second group wasn't asked to sign anything, and only 42 percent remained faithful. The final group consisted of people who were asked to sign but refused. Only 20 percent of them carried on.[3]

• **Record minor breakthroughs as you move toward your goal.** On that piece of paper headlined "Contract with Myself," you've written your goal or statement of intent. Underneath, there's plenty of room for interim progress reports.

The person who wants to fight stress and drink less might make comments such as these, which come taken from the log of one of our study participants:

9/3/91   For the first time, I completed the aerobics portion of the program, met the intensity goal, and actually felt better afterward than when I started. Usually I feel exhausted. Maybe this is what they mean by "second wind."

10/9/91   There was a major crisis at work today. Instead of following my old pattern—staying there until midnight and spinning my wheels—I told my boss, "I can't do anything about this until tomorrow when I'm fresh." I went home, did my exercise, and felt relatively calm all evening. I even got a good night's sleep.

10/25/91   I just realized that I haven't had to replenish my supply of vodka since I've been on the program. I used to go through one liter every two or three weeks. I've been nursing the same bottle now for almost two months.

11/1/91   My clothes don't fit as snugly anymore. I wasn't trying to lose weight, but if it happens, that's all right, too.

• **Exercise with a companion.** It's our experience that lonesome exercise warriors don't stay warriors for very long. That's why it's absolutely imperative that you undertake this program with a friend or mate. It's also why all the exercise drawings in this book picture two people working out together.

Companionship is a key aspect of motivation. There will be those inevitable days when you're really tempted to chuck it all. That's when a partner becomes indispensable. Having someone going through the same learning and experiential process with you will turn what may get boring at times into something you can laugh about. And on days when the program makes you feel more like crying, it's important to have a buddy to mirror your reactions and empathize.

We think success in our program is almost guaranteed if you find friends to do it with. You'll be far more likely to show up and work out if someone is waiting to walk or lift those weights with you.

• **Exercise on a regular schedule.** Partners also serve another purpose—they force you to be businesslike about exercise and set aside a specific block of time every day to do it. They help you focus on the fact that this part of your daily schedule is as important as other responsibilities. We think you'll find that if you don't exercise both with a partner and on a prearranged schedule, you'll easily find excuses not to exercise at all.

Our exercise programs are successful because we require our volunteers to come to our laboratory. If they miss an exercise session, they know we'll call them to find out why. If they miss too many, we simply ask them to leave the study. Our volunteers know they must incorporate exercise time into their daily schedule. It's as important as corporate meetings, a lunch date, or any other regular obligation.

• **Easy does it—at least in the beginning.** Far too many people embark on fitness programs only to drop out in short order because they moved ahead too fast. They tried to do too much too soon. Impatience is an unfortunate trait of human beings. It's an emotion that seizes people with full force when they get caught up with the idea of self-improvement. We find most people expect to regain in the space of two weeks the fitness they've lost slowly over the course of 30 years. This is not only foolish, it's impossible! But people try anyway and in the process become so discouraged and demoralized, they throw up their hands in despair and quit.

We'll never forget our experience with a recruit to one of our early 12-week exercise studies at Tufts. He was an older man, seemingly determined to use his participation in our study to right all the wrongs he'd ever committed against his body over the last 20 years. And he'd committed plenty of them, judging from the shape he was in.

Our man arrived toting an old pair of hiking boots. He certainly didn't need them since we would ask him to do only 15 minutes of exercise per day for the first two weeks. Never mind, he said. He was going to augment our session with a daily hike up and down the 15 floors of stairs in our research center.

Needless to say, we hid his boots and forbade him to do it. Had we allowed him to race up and down the stairs, it's guaranteed he would have either injured himself or found the whole thing so exhausting and distasteful, he would have quit exercise for good and returned to his former sedentary lifestyle, feeling all the more justified about sitting around.

We're happy to report this man is one of our success stories. When our 12-week study ended, he and his wife, another study participant, felt so fantastic and enthused about exercise training that they now plan all their vacations around hiking and sports. We still hear from them occasionally, and they claim they're more active now than they were as young adults.

One of the principal features of our program is that it's graduated to take into account the baseline fitness level of participants. It's an exercise plan that's doable. In fact, it's so easy and doable in the beginning that some of you may decide after a week that it's not worth it. Where's the challenge?

Wait. Be patient. The challenge is there. It's built into both BioAction Plans, but it won't hit you early on like a ton of bricks. Expect it to sneak up on you during the fifth or sixth week of the 16-week program for the low-fit and the third or fourth week of the 12-week program for the moderately fit. By those junctures, you'll have progressed to the point where you're exercising between 45 and 90 minutes five or six days a week.

CONTINUING TO MOTIVATE YOURSELF AFTER WE BOW OUT

Our BioAction Plans don't last forever. After the time—be it 12 or 16 weeks—is up, you'll be on your own. At that point you may once again discover that it's a little too easy to fall back into your old sedentary habits.

What follows is a list of motivators for the long haul. Return to this list every time your resolve falters—when you think you can't lift one more weight or take one more step.

- **Update your goal(s).** The goal that got you on the program in the first place is unlikely to be the one that keeps you exercising on into the future. Sustained motivation in any endeavor comes not so much from achieving a particular goal but, rather, from striving for that goal. Once a goal is reached, the average person experiences a letdown, even depression, a kind of postpartum blues. The only cure is to find a new goal, a new challenge, and start the process of mastery all over again.

We think that after you've pushed your lifestyle into higher gear via the Biomarkers Program, you'll be surprised at the vista of opportunities and new goals that open up to you. Maybe your initial exercise goal was to get an elevated blood-sugar level back down into the normal range because your father died from a diabetes-related illness. Once you feel you've got the diabetes threat under control, you may start to view exercise as a way to firm up your body and regain a more youthful physique. Or as a way to meet new friends by joining an exercise group. Or a way to test your prowess by becoming a competitive amateur athlete.

- **Learn to enjoy yourself.** Even though we don't recommend it for the reasons we've already discussed, people often begin an exercise program motivated by a sense of duty (I *must* do this!). However, they don't remain exercisers very long if they don't find inventive ways to transform what started as a duty into a source of pleasure and entertainment (I *want* to do this!). The remaining suggestions can help make exercise more enjoyable.

- **Work exercise into the fabric of your life.** While following a BioAction Plan, you should exercise on a regular schedule and with a partner. You're trying to develop a habit, so formality, structure, and camaraderie are important. While conforming to our dictates over the 12 or 16 weeks of the plan, you should also be monitoring your reactions and find out more about your body and its time-of-day and exercise preferences. Then, once you've completed our formula program, you'll be in a better position to put what you've learned into practice in a way that works for you.

As a BioAction Plan graduate, your goal is a more active lifestyle, not just an hour's worth of exercise several days a week. To that end, you must find a way to work a pattern of increased activity into your life *as seamlessly as possible*. The bigger production you make out of daily exercise, the faster you'll fail at it.

Strive to make exercise convenient as well as practical. For example, many retired people join mall-walking programs. A group of people meet regularly in the local indoor shopping mall and walk the aisles for exercise. It solves the problem of what to do in inclement weather, and it's convenient since the participants can do a little shopping before or afterward. There are numerous other possibilities: Early birds might enjoy a walk at dawn to the nearest convenience store to pick up the morning newspaper. If you live close enough to your place of employment, you might walk or bike on nice days instead of driving your car. Or instead of waiting at the nearest bus stop, how about leaving your house early enough to walk to one that's a mile away?

• **Do it your way.** As a BioAction Plan graduate, you're a free agent able to make new arrangements about when, where, and with whom you exercise. You may well find, for example, that your initial exercise partner didn't progress as fast as you— or has different exercise preferences. Fine. Move on. Find a new partner who shares your new exercise goals or has a schedule more compatible with yours.

Some people will always want to exercise at the same time in the same place each day. They're people who thrive on predictability. Other people are more adventurous and hate routine. When exercise becomes too predictable, they use boredom as the excuse to give it up.

To keep yourself motivated, there are six things you can change: (1) your stretching regimen; (2) the type of aerobic exercise you do; (3) your muscular-conditioning regimen; (4) the time of day when you work out; (5) the place; and (6) your exercise companions(s). If your whole exercise program has become too humdrum, alter one of these things. If you're really chomping at the bit, change all six at once.

• **Explore new exercise options.** By all means, tailor an exercise program to suit your own unique preferences. Experiment. Find out more about different forms of exercise and discover what you like.

Maybe you want to move beyond walking to more demanding forms of aerobic exercise and to try other people's muscular-conditioning and stretching routines. You might want to join a health club or a low-impact aerobics class to give you further incentive. Or you might want to get really serious about strength building and work out at a gym under the tutelage of a professional trainer.

Think of it this way: Now that you've restored the functional capacity you'd lost by sitting around too much, the whole wonderful world of exercise and recreation is your oyster. Just make sure that the sports or exercises that you eventually choose increase flexibility, build muscle, tax your cardiovascular system, strengthen bone, and burn fat.

• **Keep your surroundings pleasant.** In your search for the exercise(s) of your choice, keep in mind that exercising under adverse or stressful conditions is a form of self-sabotage. It's a way to give yourself an excuse for quitting. Thus, it's imperative that you always make everything about your exercise routine as appealing as possible.

Maybe you hate your swimming instructor. Or the overly chlorinated water. Or the unsanitary locker room. You don't have to endure things you don't like in the name of exercise. Simply change your health club. Or find an inviting outdoor pool. Or, more radical yet, try another sport. How about bicycling? Or volleyball?

• **Follow the dictates of the weather and the seasons.** Weather conditions that make you cringe should be avoided, too. Let's say walking is your aerobic exercise of choice. No, we don't want you to brave driving rain, snow, and sleet like the Pony Express, all in a quest for a little exercise. But we don't want you to use the weather as an excuse for skipping days, either.

Ever hear of walking in place, inside your house, where it's warm and cozy? Or how about jumping rope? Both are perfectly viable forms of aerobic exercise. Of course, some people buy treadmills for indoor walking when conditions outside prove inhospitable. It's an option, but we warn you, it's boring. The better alternative is to have a range of sports to choose from, depending on the weather and the season.

Choosing seasonal aerobic exercises is an excellent way to inject variety into your program. In summer, for example, you might swim and play tennis. In winter you could play racquetball and go cross-country skiing. To mix things up a bit, you might also change from individual sports to team sports and vice versa.

• **Take the exercise(s) of your choice seriously.** The more you know about a subject, the more interesting and stimulating it becomes.

These days there are books and special-interest magazines and newsletters about every sport imaginable. There are a battery of walking books, for instance—walking tours of city streets and back alleys as well as hiking trips on windy mountain trails. There's a whole muscular-conditioning industry you can tap into for all manner of in-depth information. By finding out why other people are so hooked on a certain form of exercise, you may become hooked, too.

• **Listen to your body.** This advice is aimed at those compulsive people who ignored one of the first motivators—Enjoy yourself! Remember, no one is holding a gun at your head. Of all the people who've ever gone on the Biomarkers Program, you don't have to be the best at it. You don't have to exercise twice as hard as our maintenance guidelines dictate and, in the process, guarantee that your muscles will throb from constant, low-level pain. Exercise discomfort is a major reason why people drop out. Reduce the frequency and intensity of your exercise to a level that affords some health rewards without making the whole experience an ordeal.

• **Perceive a lapse for what it is—a momentary interruption.** Another sign of compulsion is letting a minor setback blossom into a major catastrophe. Don't let a failure to keep exercising demoralize you so that you decide to kiss your exercise program good-bye forever. This is not only foolish at your age, it's self-destructive!

One failure does not mean that you've lost your grit and are now easy prey for all future temptation. Even the greatest athletes in the world experience times of indecision and self-doubt, so it's unfair to castigate yourself unmercifully when this happens to you. You're human, after all. So what if your self-discipline deserts you now and then? It's okay. Take comfort in the fact that occasional lapses are the rule rather than the exception.

• **Reward yourself at regular intervals.** Some people motivate themselves best through a parental script: You'll be punished if you don't meet your goal. Other people respond better to rewards for a job well done. It's the difference between wielding a buggy whip or dangling a carrot.

If you fall into the former group, we leave it to you to devise an appropriate punishment. Those in the latter group might want to design a reward system for themselves. For instance, allow yourself to buy that sweater you want after a month of loyalty to your exercise schedule. Splurge on a wonderful meal at your favorite restaurant the following month. The third month, you might want to spend a weekend at a country inn with a friend.

You get the idea. Be nice to yourself. Life is too short for austerity and constant reproaches.

• **Become a mentor to someone who can benefit from exercise the way you have.** This is a surefire way to renew your own commitment. Becoming someone else's teacher forces you to revisit the road you traveled to arrive at your present stage of advancement. As a mentor and role model, you'll also be placing yourself in a position where you must carry on. Indeed, how can you drop out if you're encouraging someone else to stick with it?

• **Consider taking part in sports competition.** We've had many a participant in our research studies become so enthralled with their favorite form of exercise that they start seeing it not only as a challenge, but as a source of pride and achievement. It's something they're good at, and they'd like the world to acknowledge it. The best way to have that happen is to enter a competition. Find out how good (for your age) you really are.

There are seniors competitions these days for every sport imaginable, from jogging and swimming to triathlon events, which combine three sports in the same contest. Your local municipal recreation department or health club is a good source of information about upcoming events. Training for competition is also an excellent way to keep yourself motivated. You've got a goal in sight and daily exercise is the way to reach it.

• **Remember the greatest motivator of them all.** Can you guess what it is? It's the fact that you can't store up the beneficial effects of exercise the way you can hoard money in a bank. Moderate exercise and/or a high degree of daily activity has to be ongoing. It has to take place almost every day. Interruptions in your usual mobility pattern will cause your fitness to plummet— *almost immediately*. This is a fact of life, and biology and you can't change it.

What this means is that you cannot double up on your exercise schedule for a month in preparation for a second month of inactivity. Like it or not, during that month of inactivity your body will once again begin to deteriorate. Your body registers positively when you exercise. It registers negatively when you don't. It's as simple and straightforward as that.

The next chapter is not intended for everyone reading this book. It's aimed strictly at serious older amateur athletes who scored exceptionally high on our self-tests in Chapter 3, people who were able-bodied exercisers long before they picked up this book.

Unless you're curious about Chapter 9's special advice to this select group of readers, you should skip ahead to Chapter 10.

FINE-TUNING AND RIGHT THINKING FOR PEOPLE ALREADY IN EXCELLENT SHAPE

———

This is the chapter for those of you who scored exceptionally high on our self-tests in Chapter 3. To be sure, the men and women in this select group didn't get themselves in such fantastic shape by lounging around. In fact, we'll wager you're all people who've been exercising for a long time. You could be marathon runners or avid swimmers or bicycle racers. From our point of view, your sport is immaterial. What matters to us is your singular dedication to physical achievement, which your age hasn't dimmed, even if your advancing years sometimes seem to be taking a toll on your abilities.

As a regular exerciser, you've probably read a book or two already about training and sports nutrition. No doubt you feel you're pretty well versed in these subjects. You're curious about what we could possibly add to your store of knowledge.

In this chapter we'd like to home in on issues of concern to the older "master athlete." Our discussion falls under two broad headings—training and nutrition. We'll give you information that will improve your performance and enable you to participate in your sport for many more satisfying years. We'll discuss what the latest research indicates about aerobic capacity and lactic acid buildup in the muscles and blood. And we'll tell you how to prepare your body, via food and appropriate rest, for that race next weekend.

We think much of what we say may surprise you.

OUR PREDICTION—THE RECORD IS ABOUT TO BE REWRITTEN

World and national records in almost all sports events are set by men and women in their 20s and early 30s. This has led to the assumption that everybody, including world-class athletes, gets slower and loses strength as they age.

We're not so sure this assumption is 100 percent valid. Let's examine it in light of the facts:

First, there is a relatively small number of middle-aged athletes competing in any sport. The reason isn't hard to fathom. It's easy for young adults to devote their lives to a competitive sport. Few have spouses or children or mortgages to worry about—at least not yet. But as they move into their 30s and 40s, their priorities tend to shift and their lives change. Making a good living becomes more pressing than winning next week's race. Even if these more mature athletes continue to compete, their lives have probably changed to the point where full-time training, 15 to 20 hours per week, is no longer possible.

Sure, middle-aged athletes don't excel in competition to the extent they did when they were younger—*but most aren't training like they did back then, either.* It's a case of comparing apples with oranges, isn't it? How can you make a blanket assertion about age sapping the ability and strength of an athlete when you're comparing a whole lot of young, 100 percent–obsessed, eager-beaver competitors with a tiny minority of older, less well trained, got-other-priorities athletes? The mere fact that the pool of young athletes is so much larger than the pool of older athletes casts grave statistical doubt on any conclusion one might attempt to draw.

Fortunately, the number of older, master's-level athletes—people who have trained all-out their whole adult life—is growing. Carlos Lopes was 39 years old when he won the Olympic gold medal in the marathon. Now, years later, he's still one of the premier runners in the world. Elite older athletes like Carlos, whose dedication to their sport has never wavered, are beginning to emerge and make their presence felt. When this group becomes substantial enough in size, we researchers will be in a far better position to judge the effect of continuous exercise on aging and physical performance. Even today there are enough master athletes like Carlos around for scientists to question seriously the old assumptions about aging and athletic achievement.

While it's true that we research scientists need more older athletes to study in order to get a clearer picture of how aging affects their performance, we need something else, too. We need to undertake longitudinal studies that follow the same group of athletes over 30 or 40 or 50 years of their careers to find out what factors are the most responsible for changes in their performance. Then we'll be able to decide if lifelong training can seriously retard those Biomarkers you learned about in Part One.

Most studies of athletic performances are cross-sectional. Young athletes' abilities are compared with those of their older counterparts, without taking into sufficient account such fundamentals as the differing training regimens of the two groups over the years or the vastly different genetic heritage of each individual. By their very nature, cross-sectional studies are biased in favor of the young. Even longitudinal studies that control for genetic differences cannot account for differences in various training techniques or intensity. Given the dearth of appropriate studies, is it any wonder we all assume that, by middle age, the career of an athlete is automatically over?

Although large-scale longitudinal studies of athletes are practically nonexistent,[1] a few investigations have followed the career of a single athlete. One study involved a male swimmer who stopped swimming competitively after college, then picked up the sport again in his late 40s.[2] Despite a 30-year lapse in swim training, *this man was able to achieve his best performance _ever_ at the grand old age of 50.*

How could this be? The researcher postulates it's due to a combination of factors—everything from latter-day improvements in swimming techniques and facilities to better training methods—not to mention the fact that the man had kept himself in good enough shape over the years to arrest any serious decline in his physiological capacity.

What we find of particular interest is the difference in the man's training regimen at ages 20 and 50. At age 20 he'd averaged about 1,500 meters a day. At age 50, with no time to lose, he swam 2,500 meters a day. Doesn't this say something about the value of training *at any age?*

Another example from several that are in the literature:

Exercise physiologist David Costill, Ph.D., analyzed laboratory data collected for 13 years on a nationally ranked distance runner whose training intensity had varied somewhat over the years.[3] Sometimes the man managed to match the intensity he'd achieved when he was 21 to 25 years old. But often his training intensity was less. Still, he managed to maintain his running ability at a relatively steady level.

In 1968, when the runner was 36 years old and training for the Olympics marathon trial, his VO_2max. (aerobic capacity) was 67.6 and his maximum heart rate was 163 beats per minute. By 1981, at age 50, his VO_2max. had declined only to 63.7 and his maximum heart rate to 155—neither being dramatic drops. At 50 he also weighed 10 pounds less. His oxygen uptake while running at the pace of 8 minutes per mile had stayed between 32 and 33 ml/kg $\times$ min^{-1} over the whole 13-year period, indicating that his running efficiency had not changed. *His performance had declined very little.*

All of which leads us to speculate that . . .

Given two athletes with similar ability and genetic heritage, it's the consistency of training that determines the extent to which one will eventually excel over the other. Their respective ages may count for very little.

Granted, this is still a hypothesis, but I think it will be borne out over the years as researchers collect more data. In the meantime, I, Bill Evans, am constantly reminded of this notion each year when I examine the fitness results of the New England Patriots football team. I'm their team exercise adviser. From the results each summer when preseason training begins, it's clear to me that a goodly number of the players do very little in the way of strenuous exercise off-season when no one's watching. From January to July—that's a long time to let your body remain fallow. Year after year, the oldest player on that team, Steve Grogan, age 37, is in the best condition with the lowest body fat.

Steve is an excellent example of the value of working hard to maintain fitness. He knows he's playing a young man's sport. If he's going to maintain the level of play required as a quarterback in the National Football League, he has to stay in tip-top shape. No matter what the season or who's around to crack the whip, Steve exercises on a regular schedule. His play shows his dedication. It's consistently excellent.

In other words, the factors that make a young athlete successful are very similar to those affecting the performance of the master athlete. Right at the top of the list is training. Consistency of effort, day in and day out. And smart eating is right up there, too.

Let's begin our discussion with training.

AEROBIC CAPACITY DOESN'T HAVE TO DECLINE SHARPLY WITH AGE

A very high aerobic capacity (VO_2max.) is essential in any endurance sport like running, cycling, swimming, soccer, ice hockey, or basketball.

You met Bill Foulk back in Chapter 2. He's the 56-year-old marathon runner whose aerobic capacity, as measured via that arduous treadmill test in our physiology lab, was the best of anyone *of any age* that we'd ever tested.

Aerobic capacity tells how much life-giving oxygen our body can utilize per minute. As such, it's the single best indicator of fitness and functional capacity there is. And Bill Foulk is living proof that, contrary to popular thinking, *aerobic capacity can be maintained at a high level at least into middle age!*

It's true, though, that aerobic capacity usually does decline in most people as they move into their 40s and 50s. That's why we included aerobic capacity as one of our Biomarkers. It's somewhat controllable, but still, by age 65, in most relatively inactive people, it is typically 30 to 40 percent smaller than in healthy young adults. Bill Foulk and his devotion to the endurance sport of marathon running prove that *what's typical is not inevitable.*

As we said earlier, aerobic capacity is a collective measurement of many physiological functions. It shows how well your lungs can take in air . . . how much oxygen is moving out of your lungs into your red blood cells . . . how well your heart pump is forcing oxygenated blood to your muscles . . . how well your capillary network is distributing this blood . . . and how productively your muscles are extracting oxygen to utilize as fuel for movement.

We also said that the latest scientific evidence now suggests that when older men and women exercise aerobically, their muscle cells—not their heart or cardiovascular system—derive the greatest benefit. Regular exercise increases their muscles' "oxidative capacity"—their muscles' ability to extract oxygen from the blood and put it to immediate, productive use.

BEYOND AEROBIC CAPACITY: THE LACTIC ACID FACTOR

While master athletes' muscles are just as responsive to training as that of their younger counterparts, the oxidative capacity of their muscles is by no means the whole story.

It's very common for two athletes to go into a race with the same aerobic capacity yet experience a widely diverging performance under competitive conditions. Leaving aside the question of mental toughness, there are physiological reasons to explain the disparity.

Research indicates that *the best single predictor of success in endurance events is not how high your aerobic capacity is, but how high a percentage of it you can sustain over a long period of time.* Some athletes can exercise at 90 to 95 percent of their maximal aerobic capacity for an extended period without becoming unduly fatigued, while others can't make it much beyond 80 percent of their capacity without experiencing crushing fatigue.

What's happening in their respective bodies that accounts for the difference?

The answer is lactic acid buildup.

Whenever we exercise, our muscles produce a substance called "lactic acid." It's a waste product of metabolism during heavy exertion. Low-intensity exertion doesn't foster lactic-acid accumulation since our muscles are able to remove the acid as fast as it's produced. However, as the intensity increases and our muscles' demand for oxygen increases, our muscle cells and blood begin to get overloaded with lactic acid.

Lactic acid in quantity is not a good thing. It increases acidity of our muscles and blood and increases our respiration and heart rate. Taken to extremes, the result is a feeling of extreme fatigue. The change in the muscles' acid level can cause a burning sensation that's almost unbearable. It's a sensation well known to sprinters. When you sprint, your muscles produce energy without the use of oxygen and accumulate lactic acid very quickly. In a matter of seconds your body is so spent and depleted that you must stop or you'll collapse from exhaustion.

Lactic acid buildup ultimately limits how long we can exercise at an all-out pace.

HIGH-INTENSITY TRAINING TO MINIMIZE LACTIC ACID BUILDUP

Fortunately there is a way to train your body to produce lactic acid at a slower rate, even though you're exerting yourself to the maximum. It's called "high-intensity interval training." The goal of this type of training is to trigger maximum production of lactic acid so your body can better learn to eliminate and neutralize it.

The table below, adapted from the work of Jack Willmore and David Costill, is a good example of "aerobic intervals" suitable for runners in training for a 10-kilometer race.

If Your Best 10-Kilometer Race Time Is . . . (in minutes: seconds)	Sprint & Rest This Many Consecutive Times	Each Repetition Consists of . . .		
		Covering a Distance of . . .	At a Pace of (in minutes: seconds) . . .	A Rest in Between No Longer Than . . .
46:00	20	400 meters	2:00	10–15 seconds
43:00	20	400 meters	1:52	10–15 seconds
40:00	20	400 meters	1:45	10–15 seconds
37:00	20	400 meters	1:37	10–15 seconds
34:00	20	400 meters	1:30	10–15 seconds

Source: Adapted from Jack H. Willmore and David L. Costill, *Training for Sport and Activity,* 3rd ed. (Dubuque, Ia.: Wm. C. Brown Publishers, 1988).

As the table indicates, interval training involves exercising at a very high intensity for a short time, followed by a very short rest, followed by more of the same. The rest period is only 10 or 15 seconds long because your body is not supposed to recover enough to clear out the lactic acid that was just formed. Rather, your blood lactic acid level should get progressively higher with each bout of maximum exertion.

All duration-sport competitors can benefit from interval training. The purpose of this type of workout is twofold:
- Interval training improves the muscles' "buffering capacity." In

short, it increases their ability to neutralize lactic acid, enabling you to tolerate higher levels of acid without becoming fatigued. You can't improve buffering capacity through regular aerobic endurance training. You can do it only through the type of sprint workout described above.

• Interval training increases the intensity at which you can run or swim or cycle or skate—or do any other form of aerobic exercise.

Interval training also builds muscle strength, as all aerobic exercise does to some extent. This, however, is not its primary purpose.

Another team for which I am the fitness adviser is the Boston Bruins Hockey team. For those of you unfamiliar with it, ice hockey involves a very high intensity, short-spurt action. The game is so exhausting that players can stay on the ice for only 60 to 80 seconds at a stretch. In that amount of time even the most fit players have accumulated such remarkably high amounts of lactic acid that they need time to recover from their overwhelming sense of fatigue. Clearly, the better conditioned the player, the faster he can recover and get back on the ice and help his teammates win the game.

The Bruins are one of the best-conditioned teams in the National Hockey League, largely because the coaches realize the value of interval training. It's an integral part of their training camp and practice sessions. Given the Bruins' consistently high standings in the league, it seems to be paying off.

Interval training can work for anybody. Figure 9-1 shows the improvement in performance of an athlete after two months of interval training. When I started counseling this athlete, he was not particularly well conditioned—probably falling into the moderately fit category. Before the special sprint training, notice how much blood lactate (lactic acid in his blood) he accumulated as he picked up speed. After the eight weeks of interval training, he could go faster, yet his blood lactate levels never reached the heights they did when he was untrained and going slower.

Before we leave this subject, we'd like to add one large cautionary footnote. High-intensity training, as you might well imagine, is not easy. In fact, it's downright unpleasant, and it can lead to injury if it's not properly supervised in the beginning.

Until you reach the point where you score Excellent on our self-tests, do not undertake interval training. Even then, the first few times you try it we advise you to do it with somebody knowledgeable about sports training and medicine. That person can clock your efforts as well as watch for any signs that you're pushing yourself beyond the limits of endurance.

QUALITY VERSUS THE QUANTITY OF TRAINING

Master athletes are particularly strong proponents of interval training because, among other things, it's highly efficient. In a relatively short time, you can get a good workout. This is not to say that lower-intensity endurance training, which is more time-consuming, can ever be dispensed with. It cannot. In fact, it's crucial for anyone whose goal is to improve, or simply maintain, aerobic capacity.

Studies on competitive long-distance runners indicate that a certain volume of training is essential to improvement.[4] As well as researchers can ascertain, the ideal training regimen seems to require an energy expenditure of some 6,000 to 10,000 calories per week. This is roughly equivalent to running 60 to 100 miles a week or swimming between 14 and 26 miles per week.

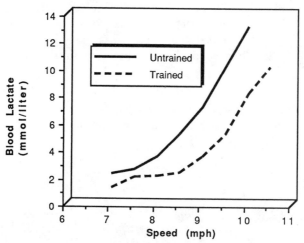

Figure 9-1 **INTERVAL TRAINING REDUCES LACTIC ACID–INDUCED FATIGUE**
As you can see in this graph, if we take a relatively untrained individual and have him or her do interval training for two months, the speed at which that person can run without accumulating high amounts of lactic acid in the blood increases substantially. This translates into better performance during competition.

Source: William J. Evans

Now we come to what we consider one of the great myths of sports competition: "There's no such thing as overtraining." Along the same lines as that famous aphorism, "You can never be too rich or too thin," some coaches and their protégés seem to believe "You can never be too prepared or too trained." We couldn't disagree more.

It's essential that all exercisers, be they elite long-distance runners or just enthusiastic amateurs working out for health reasons, incorporate rest into their training schedule. It's particularly important for the master athlete.

We've had a number of middle-aged male runners come to us boasting that they hadn't missed one day of exercise in five whole years! While they considered it an achievement, we were appalled because exercise without rest leads to overtraining. And overtraining tends to negate the benefits an exerciser has gained over many hard months of exertion.

Although overtraining remains one of those things you can't predict beforehand, you can diagnose it rather easily after the fact. Among its symptoms: [5]

- During training or competition, your performance is suddenly substandard and it stays that way for a period of time (a bad day here or there doesn't count).
- Your body feels heavy and hard to move around.
- Your appetite wanes and you lose weight.
- You experience muscle tenderness that's out of the ordinary.
- You get a series of head colds, allergic reactions, or both.
- Bouts of nausea become an occasional problem.
- You suffer from insomnia.
- You have a higher-than-usual resting heart rate, blood pressure, or both.

Your goal, then, is a challenging one—to design a training program that continues to stimulate you mentally and provide you with the optimal level of physical stress necessary for physiological improvement *without exceeding your body's abuse threshold*. Clearly, what constitutes abuse to one athlete is taken in stride by another. There are no hard and fast rules here.

While we can't say what would constitute overtraining for you, we can give you some guidelines to try to prevent it.

First of all, listen to what your body tells you and don't try to mask pain.

One 52-year-old marathon runner told us he was taking 20 aspirins a day to mask the pain of a muscle pull in his groin. Rather than give in to the pain and rest to allow sufficient time for his body to repair itself, this man was attempting to deny the pain and barrel ahead. In the process, he was constantly reinjuring his groin and almost certainly causing gastrointestinal bleeding from all the aspirin.

We do not recommend painkillers, except under doctor's orders. It's crucial for athletes to heed their body's signals. If you continue to exert yourself when you're in great pain, you can be certain you're making the problem, whatever it is, much worse. Painkillers are bad because they foster a false sense of security while you're causing more bodily damage that, after a certain point, can't be repaired or won't heal properly.

To prevent overtraining, here's another recommendation:

All exercisers—whether strictly amateur or professional-caliber—should adopt a cyclical training schedule. Mix it up. Alternate among easy, moderate, and strenuous workouts. For example, if you undertake one or two days of intense exercise, follow that up with one or two days of easy workouts, then a couple of moderate days. Rest a day or two before going back to an intense workout. Or you might make the periods longer—five or six days of peak sessions, followed by a day or two of rest and a second week of lighter sessions.

EATING TO WIN

The right diet is another strategy for warding off those heinous symptoms of overtraining. No doubt you've asked yourself, "Is there a way I can eat to achieve my best performance? Are there certain foods that will boost my ability? Is my dietary timing important?

The answer is "yes" to all three questions. Let's tackle the first two concerning what to eat:

During exercise, our muscles must produce an enormous amount of energy to propel our action. To accomplish this task, they rely on two main fuel sources—carbohydrates and fats.

As we mentioned in Part One when we discussed the Blood-Sugar Tolerance Biomarker, carbohydrates are stored in our liver and muscles as glycogen, which is nothing more than a long chain of sugar (glucose) molecules strung together. When our body puts out the emergency call for more fuel, the glycogen that's stored *in the muscles*—not in your liver—is utilized immediately. It's broken down, as needed, one sugar molecule at a time.

The amount of energy that carbohydrate contains is relatively small. Fortunately, regular exercise builds muscle tissue, enabling it to store more glycogen (carbohydrate). Still, there's a limit to the amount of glycogen your body can store. The maximum amount of energy that's stored as glycogen in a young male athlete, for example, is around 2,500 to 3,000 calories. In a young, *untrained* man the figure is closer to 2,000 calories. In a 65-year-old *untrained* man the number is even lower—some 1,500 calories.

Of course, these amounts are nothing compared to the amount of caloric energy our bodies store as fat. Even a lean long-distance runner would have almost 10,000 calories stored as fat. The trick is to get the body to utilize this fat energy before it depletes its muscle glycogen stockpile.

Someone who exercises regularly has a body conditioned to utilize both glycogen and fat calories for energy. In contrast, the sedentary person's body is less likely to burn fat calories during a sudden bout of exercise.[6] This is why we insist that *health-conscious, middle-aged people must develop a routine of regular exercise, because it's this regularity that forces their bodies' metabolism to adapt and utilize more of the calories stored as fat.*

Figure 9-2 shows where the body of an endurance athlete gets its fuel. While glycogen represents only about 12 percent of the calories an athlete burns during exercise, this relatively small figure is deceiving. *Glycogen is the most important of all the fuel sources.* Because it resides in your muscles, it's the most accessible fuel and the easiest to break down. Unfortunately, as we just pointed out, the amount of glycogen in muscles is extremely limited. And once our muscles run out of glycogen, they can no longer continue contracting to produce force even though our body still has an abundant amount of fuel stored as fat. Depleted of glycogen, your endurance flags and you find yourself among the last to finish in a long-distance event such as a marathon run, bicycle race, cross-country ski meet, and so on.

In the pie chart (figure 9-2), you'll notice that blood glucose is also an important source of exercise fuel. Basically, the glucose in your blood during exercise is in transit from your liver to your working muscles. During an extended period of exertion, this blood sugar is tapped for energy, and if the exertion continues long enough, even the liver's glycogen stores are finally mobilized.

What to Eat When

Many endurance athletes, particularly those competing in events longer than two hours, need to pay close attention to the amount and sources of the carbohydrate in their diet—and when they eat it.

As a seasoned athlete, you no doubt know about "glycogen loading." This involves eating a very high carbohydrate diet and reducing your training regimen just before the race. The idea is to increase your muscle glycogen stores by more than 50 percent to insure that you won't become glycogen-depleted before you cross that finish line.

Suppose you're preparing for a long endurance event next Saturday. Here's what we advise you to do five days beforehand:

Tuesday. Undertake a long, intense workout with a twin goal —to deplete yourself of muscle glycogen and to increase your muscles' sensitivity to the hormone insulin. Insulin-sensitive muscles are muscles that will pull more glucose out of the bloodstream to transform into glycogen.

That night, eat a meal that's made up of 60 to 70 percent carbohydrate, which is an unusually high amount. A good menu

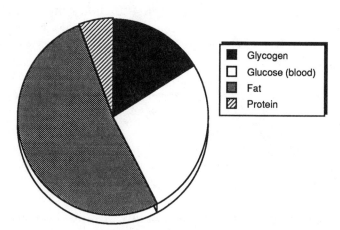

Figure 9-2 **THE CONTRIBUTION OF DIFFERENT FUELS DURING A TWO-HOUR ENDURANCE EXERCISE SESSION**
Over time, a person's body adapts to regular exercise by utilizing a greater range of caloric fuel sources. This pie chart shows the caloric energy sources (the type and amount) that a well-trained athlete's body burns during a long bout of moderate-intensity exercise.

Source: Wm. J. Evans and Virginia A. Hughes, "Dietary Carbohydrates and Endurance Exercise," _American Journal of Clinical Nutrition_ 41 (May 1985): 1146–54.

might be salad, spaghetti with tomato sauce, bread, fruit juice, and, for dessert, ice milk or angel food cake.

Wednesday. From now until the race on Saturday, begin to reduce your training schedule by at least one third and continue to consume the same high-carbohydrate diet we described above. That means, over a 24-hour period, you'll probably be consuming between 2,000 and 3,000 kcals (or calories, in lay parlance) of carbohydrate.

From now until the race, do not drink any *artificially sweetened* drinks. Do drink plenty of fruit juice, which contains a natural form of sugar. Sugar is carbohydrate, after all.

Thursday. Reduce, or taper, your exercise session even more than you did yesterday. Continue your carbohydrate-heavy diet.

Friday. REST! REST! REST! If you can't take the day off and relax, at the very least refrain from any formal exercise. Continue your carbo-loading and drink enough water or juice throughout the day to keep your urine from turning dark. Before you go to bed, drink an extra glass of water.

Saturday—the day of the race. If the event is at midday, eat an extremely high carbohydrate breakfast, such as waffles with syrup and orange juice. The purpose is to replace the glycogen in your liver. You need to restock your liver's storehouse after the extensive pilferage that takes place every night while your body rests and fasts.

Suppose it's an early-morning race, say, 8 A.M. Then you've got a couple of choices. You can get up at 4 or 5 A.M. and drink a large glass of juice—but do not eat much for breakfast. Or you can get up later, but *do not eat or drink a high-sugar or -carbohydrate breakfast;* maybe settle for some toast with jelly. Why? Eating pancakes and syrup with juice at 6:30 or 7:00 A.M. will cause an insulin surge, then a hypoglycemic reaction. We'll explain this more fully in a moment.

No matter what hour the race begins, do not consume any highly sweetened—natural or otherwise—drink for two hours prior to great exertion.

This is extremely important and a mistake many amateur athletes make.

Ignore this advice and you may share the fate of a 53-year-old rower who consulted with us because he couldn't understand why he felt dizzy at the end of every race. On several occasions he'd even passed out. His doctor thought it might be a precipitous drop in blood pressure following short, high-intensity bouts of exercise. But, after testing, that avenue of inquiry proved to be a dead-end street.

We suspected something else. We took a blood sample at the end of a race to evaluate his glucose level. Sure enough, his postrace blood glucose level was about 40 mg percent; the normal is about 90 mg percent. No wonder he was seeing the world turn before his eyes. The culprit was the quart of Gatorade he habitually drank an hour before every race. Many so-called sports drinks contain a lot of sugar along with the helpful electrolytes. Gatorade contains 6 percent sugar. Coke is about 11 percent sugar. The sugar in the drink caused a marked increase in his glucose levels. (Any very sweet food would have the same effect.) After this steep rise in his glucose levels, the six high-intensity minutes he spent crewing had the opposite effect—causing his blood glucose levels to reverse and plummet. End result: pronounced hypoglycemia. (Figure 9-3 illustrates what happens to blood glucose in such situations.)

Hypoglycemia makes you feel terrible. Symptoms range from dizziness to nausea and headache. Hypoglycemia makes exercise seem far more intense than it is and causes premature exhaustion.

Our rower isn't alone. A lot of athletes make the mistake of loading up on sugar—eating candy, drinking something ultra-sweet—before intense exercise. It's one of the worst things you can do.

Sugar, being a carbohydrate, has the effect of immediately elevating your insulin level. Exercise magnifies the stimulative effects of insulin, and therein lies the problem. High blood insulin fosters the fast entry of glucose into muscle cells while it's simultaneously shutting down the production of glucose by the liver. End result: an extremely low blood glucose level, triggering hypoglycemia. ("Hypo" means low.) Since the only fuel for the brain is glucose, it's easy to see why you'd experience these symptoms.

From this you might infer that sugar is to be avoided as a carbohydrate source in favor of complex carbohydrates. It's to be avoided *two hours before exercise,* to be sure. But, other than that, as a serious athlete, feel free to indulge yourself if sugar is your passion—and bread, rice, and pasta are not. Studies show that 100 calories' worth of jelly beans will replace as much glycogen in your body as 100 calories' worth of potatoes. Complex carbohydrates are recommended over simple sugars, however, for another reason. Complex carbohydrates are dense calories. They come packaged with beneficial vitamins and minerals, which you don't get much of in most very sweet foods.

Training for Short Duration Events

Training for shorter events isn't much different from what we described above for longer races.

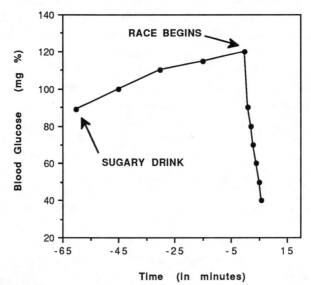

Time (in minutes)

Figure 9-3 **AVOID SUGAR TWO HOURS BEFORE INTENSE EXERTION**

Don't make the mistake of loading up on sugar just before exercise and then wondering why your performance suffers. Your performance is substandard because you've caused your body to go into hypoglycemic shock.

This graph shows what happens to the blood glucose levels of people who drink a sugary beverage about an hour before a race. Before they drink it, their blood sugar is normal, about 90 mg percent. Moments after the sugary fluid slides into their stomach, their blood glucose begins to rise to abnormal levels. By the time the race begins, it's up to 120 mg percent. Only five minutes into the race, it's done a complete reversal, plunging to a radically subnormal 40 mg percent.

The explanation is simple: Sugar immediately elevates the body's insulin level, and exercise magnifies insulin's stimulative effects. High concentrations of insulin foster the fast entry of glucose into muscle cells while the body is simultaneously halting the liver's glucose production. End result: extremely low blood glucose levels. Robbed of its fuel, the brain rebels with a handful of unpleasant symptoms—dizziness, nausea, headache, and premature exhaustion.

Source: William J. Evans

In general, we've found that sprint-type athletes make the mistake of not resting enough immediately before the big event. Rest is a critically important part of anyone's regular training schedule. *At least one day a week of rest is important for any exerciser —and certainly for the master athlete.*

Another mistake that's common involves confusion about what constitutes a high-carbohydrate diet. Many runners and athletes come to us thinking that cookies and doughnuts are high-carbohydrate foods. Anyone who has ever made a chocolate-chip cookie or any kind of cake knows full well how much butter and shortening goes into the ingredient mix. In truth, these are high-*fat* foods. And high fat intake is not helpful for the athlete—or anyone, for that matter.

The Boston Bruins learned this lesson the hard way a few years ago. Some of the players were complaining of feeling tired all the time. Add to this lassitude the very real fatiguing demands of the game, and we had players on our hands who couldn't compete at the intensity required to prevail over their opponents.

We ruled out a number of possible causes. It was clear they were training enough. The team was playing two or three games a week and practicing on the off days. But were they training too much?

I finally decided that nutrition and inadequate rest could be the problem and asked them to write down everything they'd eaten for three days. Sure enough, the players complaining of the fatigue had an extremely high fat, low carbohydrate diet. As a consequence, they were using up a substantial amount of their muscle glycogen stores every day. Hockey is played at an extremely high intensity—above 150 percent of VO_2max. Because of this, hockey players have the greatest rate of glycogen use of athletes in any sport. Their heavy training schedule wasn't allowing their bodies adequate time to make glycogen from whatever small amount of carbohydrates there was in their diet.

Needless to say, we immediately revised the team's training schedule and passed out diet plans for players to follow.

What About Protein, the Body-Builder Nutrient?

Our studies show that the need for protein in the diet is greater in athletes *of any age*[7] than the current adult daily Recommended Dietary Allowance (USRDA) of .8 grams of protein per kilogram of body weight. We've found that athletes need at least .9 grams of protein per kilogram of body weight per day—and maybe as much as 1.3 grams. Inadequate protein may also prevent you from getting the maximum benefit from your training.

It's true that people in most industrialized countries have plenty of protein in their diets. On the other hand, athletes tend to be especially health-conscious, and some may mistakenly believe that all the talk about red meat and dairy products being high-fat foods, hence bad, applies to them. Unfortunately we cannot make a definitive statement in this regard because the data from numerous studies targeting the effect of exercise on blood cholesterol are not conclusive. The data do show that if you lose weight when you train, your total cholesterol will go down. However, if you maintain your weight during training, there will be very little effect on total cholesterol.

In short, don't rob your body of protein. Animal proteins are complete. They contain all the essential building blocks (amino acids) to synthesize protein in your body. If you're a vegetarian, soy protein (such as tofu) must become a staple in your diet. It, too, is a complete protein. Incidentally, there is no evidence that regular exercise increases the requirement for specific vitamins or minerals, so do not feel that, as a serious athlete, you have to take supplements.

Fluids—Keep on Drinking

We've emphasized the need for fluid replacement throughout this book. We come back to it once again because it's the older athlete who is at greatest risk of dehydration.

As we pointed out in Part One, age diminishes our thirst and kidney function. These changes can lead to a state of chronic dehydration in older people. Here are some tips to insure you're not among this group:

- Weigh yourself on a reliable scale before and after you exercise. Drink enough water—or other nonalcoholic beverage—to bring your weight back to preexercise levels.
- If your urine is dark, it means you're dehydrated. Immediately drink enough fluid to bring your weight back to preexercise levels.
- Drink alcohol in moderation—and *never, ever before exercise*. Alcohol is bad before exertion because it tends to increase urination, hence deplete your body even faster of fluids.
- Be careful about your intake of coffee and other caffeinated bev-

erages. Every coffee drinker knows that it increases urine production and thus has a dehydrating effect.

Many athletes use caffeine to hype themselves up and enhance performance. Studies show that caffeine in the body increases the use of fats during exercise and thus decreases the body's reliance on the muscles' precious glycogen stores. A large number of studies now indicate that caffeine can increase the time it takes to become exhausted and can also alter a person's perception of exertion. In other words, caffeine makes you think you're not exercising as hard as you really are. In terms of the Borg Scale we reproduced in Chapter 4, with coffee in their system, exercisers tend to perceive level 17 ("very hard") as a 15 ("hard") or a 13 ("somewhat hard").

As a serious athlete, you should be aware that the U.S. and International Olympic Committee have identified caffeine as a potential performance-enhancing drug and have, therefore, banned its high-dose consumption during competition.

BRINGING YOUR EXPECTATIONS IN LINE WITH REALITY

We've just given you pages of advice to improve your athletic performance. By doing so, are we raising your expectations unrealistically ? In short, just how trainable is the older athlete?

Chronological age will eventually take its toll, you can be sure of that. Maybe you're 63 and you haven't noticed much change yet. Your best friend, whom you've been training with for years, is only 58. Already he or she seems to be losing it.

It just goes to show that everybody is different. Everybody is unique with a different genetic makeup and mental and emotional apparatus. (Yes, the latter also figures into athletic performance.) It's hard to predict when a person will begin to notice that his body no longer responds to training the way it once did.

While the distinction between master athletes and young athletes, particularly in endurance sports, is growing fuzzier, at a certain point—a certain age that varies with each and every individual—you'll find that neither the volume of training, nor high-intensity interval workouts, nor eating a high-carbohydrate diet at strategic times will keep you competitive with younger athletes who, perhaps, aren't even doing as much as you to stay fit. That's the reality you face. Someday.

There are several reasons why, in the long run, chronology is destiny. Not even master athletes can escape those inevitable Biomarkers.

There are two Biomarkers that master athletes find the most disturbing because they interfere the most with peak performance.

• The first involves strength.

Yes, you can increase the strength and size of your muscles at any age by making each muscle cell grow bigger. What you can't do through exercise is increase the number of muscle cells in your body.

For some inexplicable reason science has yet to fathom, people lose muscle cells as they age. And the muscle cells most affected are those that compose the fast-twitch fibers we talked about in Chapter 2. Fast-twitch fibers are what we call upon for power, to lift heavy objects or to do high-intensity, sprint-type exercise. (Slow-twitch fibers are those necessary for posture and low-intensity movement.) A decline in fast-twitch fibers means your quickness, speed, and ability to outkick the competition will eventually be affected by age.

Fortunately, depending on your sport, there are strategies you can employ to compensate for this loss. Endurance athletes are in a particularly good position to deploy them.

Take the case of Alberto Salazar, the middle-aged champion marathon runner. In truth, Alberto has run at an advantage his whole life. When he visited our lab for tests, we discovered, through a muscle biopsy, that he has the highest percentage of slow-twitch fibers we've ever seen in anyone. His muscles were 98 percent slow-twitch fibers, while the average for elite long-distance runners is about 80 percent. Mother Nature is responsible, and there isn't anything Alberto, or anyone else, can do about it.

Here is the race strategy for an Alberto Salazar—or an older runner with fewer fast-twitch fibers—to use:

When you're running a mid- or long-distance race—a 5- or 10-kilometer race or longer—your brain "recruits" your body's slow-twitch fibers almost exclusively. While the slow-twitch fibers are mobilized for action, your fast-twitch fibers are essentially resting. It's at the end of the race when you see you've got a few contenders ahead of you that you consciously have to turn up the steam and enlist the aid of those fast-twitch fibers for one last all-out sprint to the finish. Clearly, if you're Alberto Salazar, you're in no position to rely on a "kick" at the race's end to put you past your adversaries. Therefore, his strategy—and yours— is to set a pace throughout the race that's fast enough to make a kick at the finish unnecessary.

Many older runners, swimmers, and cyclists find it necessary to give up short races in favor of longer ones, simply because their speed has diminished much more than their endurance. Indeed, sprinters seem to peak around age 19 or 20. That makes sense since fast-twitch muscle fiber cells, from everything that we scientists now know, seem to start disappearing around age 20.

• The other Biomarker that will noticeably curtail your performance involves the heart.

As we age, our hearts, even under maximum stress, beat slower and slower. This is because our hearts grow less responsive to norepinephrine, the hormone that exertion triggers to stimulate cardiac output. Lower maximal cardiac output—the amount of blood our hearts can pump under the most dire conditions— means that less oxygen reaches the working muscles when they need it most. It also means a lower aerobic capacity.

In older sedentary people, a smaller heart due to inactivity also contributes to a lower cardiac output. As a master athlete, this doesn't apply to you. From years of working out, your heart should maintain its size or be somewhat enlarged, if anything. Still, the evidence to date seems to indicate that neither your heart muscle's robust size nor intense exercise training can turn the tide against that inevitable decline in your maximal cardiac output.

However, having said this, we must concede that these observations are based on comparing young people's hearts with older people's hearts. To date, no researcher has ever followed lifelong practitioners of regular exercise to find out if these age-related cardiac changes are equally true of them or if any cardiac decline that does occur in master athletes perhaps takes place at a later age —and/or much more gradually.

The marathon runner we met earlier in this chapter gives us reason to think these suppositions have validity. When David Costill followed that 36-year-old male runner from the time of his 1968 Olympic trial to age 50, he found *no significant age-related change in aerobic capacity and a very small reduction in the man's maximal heart rate.* Most important of all, the runner's performance showed little sign of age.

Part IV

STRETCHING YOUR HEALTH SPAN VIA DIETARY CHANGE

CHAPTER
10

THE CHALLENGES FACING THE HEALTH-CONSCIOUS EATER

——

It may seem that we've been overemphasizing the contribution that exercise makes to maintaining our 10 Biomarkers of vitality. This is not to discount the importance of a good diet. In this chapter we'll focus on diet and discuss the adverse impact that aging can have on your eating habits and nutritional status. Then we'll describe what you can do as you grow older to blunt these age-related challenges.

However, the message we want you to glean from this chapter is that *diet and exercise are interrelated*. Regular exercise—plus a diet geared toward specific age-related needs—is a powerful BioAction duo. It increases your chances of staying well longer, retarding the progression of those inevitable Biomarkers, and even postponing the onset of some chronic diseases. In short, it's a prescription for a lengthy health span and a shorter Disability Span.

Take our word for it, a good diet and little or no exercise won't help much to slow the aging process. Nor will avid exercise and a poor diet. If you want to maintain the vitality of those dynamic Biomarkers, the answer is allegiance to the dietary principles we outline in this chapter *plus* regular, moderate exercise. The two go hand in hand, as you're about to learn.

EATING FOR ACTION AS YOU AGE GRACEFULLY

Let's start with basics: What's a sensible diet for adults of all ages?

Our recommendations aren't going to startle you. Nor will you find them particularly complicated or radical. We simply want you to eat sensibly and moderately in line with the amount of exercise you get. In essence, we want you to eat less like a typical American, who takes his or her diet cues from television commercials extolling fast, convenience foods that are often high in fat, and more like a person who is conscious of the importance of good nutrition and appreciates Mother Nature's bounty of vegetables, fruits, and grains.

Our healthful menu dictates are these (see figure 10-1):

• Complex carbohydrates should account for up to 60 percent of your total daily caloric intake. Good sources are fresh fruits and vegetables, whole-grain breads and cereals, and such starches as potatoes, rice, and pasta.

• Protein should be limited to a minimum of 10 percent and a maximum of 20 percent of your daily diet. You might think a strength-building program such as ours would warrant a high protein intake to maintain muscle mass. It does not. First, Americans and people in the developed countries of the West tend to have more than adequate protein in their diet already, being the avid meat eaters that we are. Second, *too much protein in older people can place undue stress on the kidneys, whose function declines with age.*

• Fat should not exceed 30 percent of daily calories. Less than a third of that should be saturated, which is fat derived mostly from animal sources. Saturated fat can generally be reduced by eating fewer fatty meats, less butter, and less hydrogenated and tropical oils. By using the right kinds of oils—monosaturates and polyunsaturates—in your spreads and salad dressings and in preparing foods, you can improve the mix of saturated and unsaturated fat in your diet.

Figure 10-1 **EATING ACTION PLAN FOR MATURE ADULTS**

This pie chart gives a graphic depiction of the ideal mix of nutrients in the diet of a mature adult, aged 45 and older. Complex carbohydrates should account for some 60% of total daily caloric intake. Fat, at the very most, should account for 30%; and you should eat as little saturated fat as possible. Protein should make up 10 to 20%.

Eating in a healthy fashion doesn't have to be complicated. In general, just try to eat more fresh vegetables, fruits, cereals, and legumes. Limit your consumption of red meats and cold cuts, substituting poultry without skin and fish as often as possible. Avoid fried and other fatty foods. And substitute low- or nonfat dairy products for their whole-milk counterparts.

Daily Diet

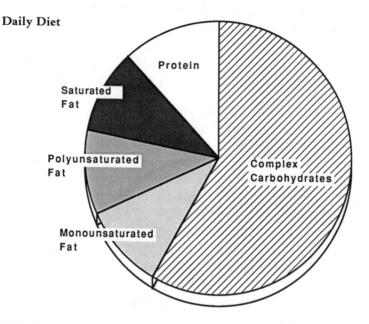

Carbohydrates

Good Sources:

Fresh fruits
Fresh vegetables
Legumes (beans, lentils)

Whole-grain breads & cereals
Starches (potatoes, rice, pasta)

Fat

Monounsaturated:
Olive, canola, and peanut oils; avocados; some nuts
Polyunsaturated:
Safflower, corn, sunflower oils; margarine, mayonnaise; most nuts
Saturated:
Animal sources such as meat and dairy products; tropical vegetable oils
   (coconut, palm, etc.); hydrogenated vegetable oils.

Protein

Dried beans and peas (also
   considered complex
   carbohydrates)
fish

poultry
lean red meats
low-fat dairy products
eggs

This is all well and good, but many of you may not be that familiar with which foods are high in carbohydrate, protein, and fat—or the essential vitamins and minerals you need in your diet. We devised the following Food Group lists and Daily Calorie Goal charts with several purposes in mind. First, they give you a game plan for deciding what you should eat every day in a way that apportions calories properly among the three major classes of nutrients. In addition, they afford a panoramic view of your dietary choices in all their variety and splendor.

FOOD GROUPS

To give you maximum flexibility in designing a healthful daily menu, we've put together the following lists of foods grouped in nine categories: (1) starches; (2) meat and meat alternatives; (3) dairy products; (4) vegetables; (5) fruit; (6) fats; (7) free food; (8) combination foods; and (9) foods to eat only occasionally. This table is to be used in conjunction with our five Daily Calorie Goal charts covering 1,200, 1,500, 2,000, 2,500, and 3,000 calories a day. The Daily Calorie Goal charts, which appear a little later in this chapter, tell how many servings from each of the food groups to eat every day in order to stay within a designated calorie limit.

BEING SELECTIVE ABOUT WHAT YOU EAT— AS WELL AS HOW MUCH

The goal of people who are serious about preserving the vitality of their 10 Biomarkers is to tip their body's balance in the direction of more muscle and less fat tissue. On our BioAction Exercise Plan, you'll be raising the calories you expend every day and your metabolic rate. Thus, even if you merely keep what you eat constant—eating the same foods in the same amounts as you did before you ever picked up this book—you should gain muscle and lose fat. The problem is that your newfound exercise regimen will, in all likelihood, increase your appetite. A hardy appetite is fine *provided* you're sensible and selective about the foods you eat.

To help you in this regard, we've devised the five Daily Calorie Goal Charts below. They're sample daily menus, falling within caloric ranges appropriate to people who are middle-aged and older. These charts, when used in conjunction with the Food

(continued on page 231)

Starches
The serving amounts listed contain approximately 70 calories.

Breads

White, ★whole-wheat, or rye—1 slice
★Whole-grain bagel—½
Biscuit or muffin (2 inches in diameter)—1
Bun (hamburger or hot dog)—½ provided 8 buns constitute 1 pound
★Cornbread (1½-inch cube)—1
English muffin—½

Crackers

Graham (2½-inch squares)—2
Melba toast—4
Oyster crackers—20 or ½ cup
Saltines—5
Round, thin—6
★Ry-Krisp—3
★Tortilla (6 inches in diameter)—1
Wheat crackers—5

Cereals/Rice/Pasta

★Whole-grain hot cereal—½ cup
★Dry bran flakes—⅔ cup
Dry puffed cereal—1½ cup
★Cooked brown rice—½ cup
Cooked white rice or grits—½ cup
★*Enriched,* cooked spaghetti, macaroni, noodles, or other pasta—½ cup

Vegetables

Dried, cooked ★beans (lima, navy, kidney) or peas (blackeyed, split, etc.)—½ cup
★Baked beans (no pork)—¼ cup
Corn—⅓ cup or ½ medium ear
White potatoes—½ cup or 1 small
★Sweet potatoes or yams—¼ cup
Popped popcorn (no butter)—1 cup

★ The stars next to some foods indicate they are rich in nutrients that older people need to meet their special dietary requirements, a subject we discuss at length at the end of this chapter.

Meat and Meat Alternatives

The following low-calorie, *lean* meats and meat substitutes, in the serving amounts listed, contain approximately 50 calories. Choose cuts of meat that are *not* marbled with fat; trim off any visible fat around the periphery. Equally important, do not add fat during the cooking process. Baking, broiling, and poaching are the preferred cooking methods.

Dried, chipped beef—1 ounce

LEAN-ONLY cooked *beef, lamb, pork, ham, veal—1 ounce

*Liver—1 ounce

Cooked, *skinned* poultry—1 ounce

Any fish except those listed below—1 ounce

Crab—¼ cup

*Clams, *shrimp, or *oysters—5 medium-size

Scallops—1 large if 12 constitute 1 pound

Canned tuna packed in water—¼ cup

*Canned pink salmon—¼ cup

*Cottage cheese—¼ cup

Medium-Fat Foods: The following meats and meat substitutes are higher in calories, containing approximately 73 calories in the serving amounts listed. Thus, you should eat them sparingly.

MEDIUM-FAT ONLY, cooked beef, lamb, pork, veal—1 ounce

Duck—1 ounce

Goose—1 ounce

Poultry with skin—1 ounce

Cold cuts—1 ounce

Frankfurters—1 when 8–9 constitute 1 pound

Vienna sausages—2

Cheese (brick, cheddar, Roquefort, Swiss, processed, etc.)—1 ounce

Whole egg—1

Red salmon, canned or smoked—¼ cup

Sardines—3 medium

Canned tuna packed in oil—¼ cup

Peanut butter—2 teaspoons

Dairy Products

The following items, in the serving amounts listed, contain approximately 12 grams of carbohydrate, 8 grams of protein, traces of fat, and 80 calories.

*Nonfat milk—⅓ cup dry; 1 cup when liquefied

*Fat-free buttermilk—
    1 cup

*Plain yogurt made with *nonfat* milk—1 cup

If you eat the following, in exchange you must omit one fat serving:
Lowfat milk—1 cup

*Plain yogurt made with *lowfat* milk—1 cup

If you eat the following, in exchange you must omit two fat servings:
Whole milk—1 cup

*Carnation evaporated milk—½ cup

*Buttermilk made from whole milk—1 cup

Vegetables

Do *not* prepare these vegetables by adding cream, sugar or honey, or fat. Serving size: ½ cup cooked; 1 cup raw—equivalent to about 36 calories.

★Asparagus	Cucumbers	Radishes
Bean sprouts	★Dandelion greens	Sauerkraut
★Beet greens	Escarole	Onions
★Broccoli	Eggplant	Turnips
★Brussels sprouts	Green beans	★Spinach
Cabbage (any kind)	Beets	Summer squash
Cauliflower	Winter squash	Tomatoes
Celery	★Kale	★Turnip greens
Carrots	Lettuce (any kind)	Watercress
Green peas	Mushrooms	Wax beans
★Chard	★Mustard greens	Artichokes
★Chicory	Okra	Pumpkin
★Collard greens	★Peppers (green or red)	Rutabaga

Fruit

The following items, *eaten without sugar or syrup,* contain approximately 40 calories in the serving amounts indicated. These fruits can be fresh, cooked, dried, frozen, or canned.

Apple (medium)—½	Fresh peach (medium)—1
Applesauce—½ cup	Canned peaches—½ cup
Fresh apricots (medium)—2	Dried peach halves—2
Dried apricot halves—4	Fresh pear (small)—1
★Banana (small)—½	Canned pears—½ cup
Blackberries—1 cup	Dried pear halves—2
Blueberries—⅔ cup	Pineapple—½ cup
Boysenberries—1 cup	Fresh plums (medium)—2
★Cantaloupe (medium)—¼	Prunes—2
Cherries (large)—10	Raisins—2 tablespoons
Dates—2	Raspberries—1 cup
Fresh figs (large)—1	★Strawberries—1 cup
Dried figs—1	★Tangerine (large)—1
Canned fruit cocktail—½ cup	Watermelon cubes—1 cup
★Grapefruit (small)—½	*Juices:*
Grapes—12	Apple juice—⅓ cup
Honeydew melon—⅛	Grape juice—¼ cup
★Mango (small)—½	★Grapefruit juice—½ cup
Nectarine (small)—1	★Orange juice—⅓ cup
★Orange (small)—1	Pineapple juice—½ cup
★Papaya (medium)—⅓	Prune juice—¼ cup

Fats

The serving amounts listed contain approximately 45 calories.

★Avocado (4 inches in diameter)—⅛

Bacon that's cooked until crisp—1 slice

Butter or margarine—1 teaspoon

Sour cream—2 teaspoons

Cream cheese—1 tablespoon

Nuts—6 small ★Almonds, walnuts, cashews, etc.

French salad dressing—1 tablespoon

★Mayonnaise—1 teaspoon

Roquefort salad dressing—2 teaspoons

Thousand Island salad dressing—2 teaspoons

★Oil—1 teaspoon

Olives (small)—5

Free Foods

This disparate assortment of "free foods" includes condiments, seasonings, and beverages that contain less than 20 calories per serving. We haven't specified serving size because you can eat or drink as much as you want of these items _provided_ you don't add cream, sugar or honey, or fat. Also, be sparing with the salt if you must use it at all.

Coffee or tea	Parsley	_Desserts and Fruits:_
Clear broth	Herbs	Cranberries
Bouillon	Spices	★Lemons
Beverage (artificially sweetened, containing less than 5 calories per 8 ounces)	Seasonings	Gelatin (unsweetened or artificially sweetened)
	Flavorings	_Juices:_
	Vinegar	★Lemon juice
Jelly (artificially sweetened)	Mustard	Tomato juice
	Horseradish	Vegetable juice
Sugar substitute	Salad dressing (dietetic)	

Combination Foods[1]

Combination foods do not fit neatly into any one of the above categories. Also, unless you make a combination food yourself—or it comes with the ingredients listed on the package—it's impossible to tell exactly what's in it. Below we list some popular combination foods. The third column indicates how these foods should be integrated into the food lists above. For example, if you have canned spaghetti and meatballs for dinner, you must assume this one dish is equivalent to two choices from the Starches list, one selection from the Medium-Fat Meats list, and one selection from the Fats list. As you can see, when you're watching the number of calories you eat, you can make life much easier by eating "pure" rather than "combination" foods as much as possible.

Food	Serving Amount	Equivalent to:
Homemade casseroles	1 cup or 8 ounces	2 items on the Starches list 2 items on the Medium-Fat Meats list 1 item on the Fats list 350 calories
Cheese pizza with a thin crust†	¼ of a 15-ounce (or 10-inch) pizza	2 items on the Starches list 1 item on the Medium-Fat Meats list 1 item on the Fats list 500 calories
Chili with beans‡ (commercial)	1 cup or 8 ounces	2 items on the Starches list 2 items on the Medium-Fat Meats list 2 items on the Fats list 250 calories
Chow mein‡ (without noodles or rice)	2 cups or 16 ounces	1 item on the Starches list 2 items on the Vegetables list 2 items on the Lean Meats list 500 calories
Macaroni and cheese†	1 cup or 8 ounces	2 items on the Starches list 1 item on the Medium-Fat list 2 items on the Fats list 250 calories
Soup Bean‡	1 cup or 8 ounces	1 item on the Starches list 1 item on the Vegetables list 1 item on the Lean Meats list 170 calories
Chunky, all varieties†	10¾-ounce can	1 item on the Starches list 1 item on the Vegetables list 1 item on the Medium-Fat Meats list 200 calories

Food	Serving Amount	Equivalent to:
Creamed† (made with water, not real cream or milk)	1 cup or 8 ounces	1 item on the Starches list 1 item on the Fats list 70 calories
Vegetable† or broth†	1 cup or 8 ounces	1 item on the Starches list 16 calories
Canned spaghetti and meatballs†	1 cup or 8 ounces	2 items on the Starches list 1 item on the Medium-Fat Meats list 1 item on the Fats list 250 calories
Sugar-free pudding (made with skim milk)	½ cup	1 item on the Starches list 100 calories
Dried beans,‡ peas,‡ or lentils‡ as a meat substitute	1 cup (cooked)	2 items on the Starches list 1 item on the Lean Meats list 230 calories

† 400 mg. or more of sodium per serving

‡ 3 grams or more of fiber per serving

Foods to Eat Only Occasionally[2]

The following foods fall into the special-treats category. They are essentially indulgence foods for people with a hankering for sweets. You'll notice they are to be eaten in small amounts because they are concentrated sources of simple carbohydrate that pack considerable calories.

Food	Serving Amount	Equivalent to:
Angel food cake	1/12 cake	2 items on the Starches list
Cake *with no icing*	1/12 cake or a 3-inch square	2 items on the Starches list 2 items on the Fats list
Cookies	2 small ones that are each about 1¾ inches in diameter	1 item on the Starches list 1 item on the Fats list
Frozen fruit yoghurt	1/3 cup	1 item on the Starches list
Gingersnaps	3	1 item on the Starches list
Granola	¼ cup	1 item on the Starches list 1 item on the Fats list
Granola bars	1 small	1 item on the Starches list 1 item on the Fats list
Ice cream, any flavor	½ cup	1 item on the Starches list 2 items on the Fats list
Ice milk, any flavor	½ cup	1 item on the Starches list 1 item on the Fats list
Sherbet, any flavor	¼ cup	1 item on the Starches list
Snack chips,† all varieties	1 ounce	1 item on the Starches list 2 items on the Fats list
Vanilla wafers	6 small	1 item on the Starches list 1 item on the Fats list

† If you eat more than one serving, be aware these foods have 400 mg. or more of salt.

Groups lists above, will enable you to design a daily diet that will appease your appetite while you're shedding fat and gaining health-promoting lean-body mass.

A PRECAUTIONARY NOTE ABOUT DIETARY FAT

In looking over our sample daily menus, you'll notice we're extremely stingy about fat. For good reason:

All calories are not created equal. In contrast to carbohydrate and protein calories, the fat calories in food are easily converted into body fat tissue.

(continued on page 236)

Daily Calorie Goal Chart: 3,000 Calories

Food Group	Number of Servings
Starches	16
Meat/Meat Alternatives	11
Dairy	3
Vegetables	4
Fruit	9
Fats	12
Free Foods	as desired

Daily Menu Example

These foods were chosen from our Food Groups lists appearing earlier in this chapter.

Breakfast

1 serving Dairy	1 cup skim milk for cereal and coffee
3 servings Starches	⅔ cup bran flakes
	2 slices whole-wheat toast
3 servings Fruit	½ cup prune juice; ½ banana for cereal
3 servings Fat	3 teaspoons margarine
1 Free Food	1 cup coffee

Lunch

3 servings Starches	2 slices whole-wheat bread
	¾ ounce unsalted pretzels
4 servings Meat	4 ounces sliced turkey
3 servings Fat	3 teaspoons mayonnaise
2 servings Vegetables	1 cup lettuce and tomato
	4 ounces V-8 vegetable juice
2 servings Fruit	1 medium apple
1 Free Food	iced tea with lemon

Snack

2 servings Starches	6 Ry-Krisp crackers
1 serving Meat	2 teaspoons peanut butter
1 serving Dairy	1 cup *nonfat* yogurt

Dinner

6 servings Meat	6 ounces broiled red snapper
2 servings Vegetables	1 cup steamed broccoli
3 servings Starches	1 cup rice; 1 dinner roll
3 servings Fat	3 teaspoons margarine
1 serving Dairy	1 cup skim milk
2 servings Fruit	4 apricot halves (in their own juice)

Snack

5 servings Starches	5 cups air-popped popcorn
3 servings Fat	3 teaspoons melted margarine
2 servings Fruit	1 cup orange juice

Daily Calorie Goal Chart: 2,500 Calories

Food Group	Number of Servings
Starches	13
Meat/Meat Alternatives	8
Dairy	3
Vegetables	3
Fruit	9
Fats	9
Free Foods	as desired

Daily Menu Example

These foods were chosen from our Food Groups lists appearing earlier in this chapter.

Breakfast

1 serving Dairy	1 cup skim milk for cereal and coffee
3 servings Starches	⅔ cup bran flakes
	2 slices whole-wheat toast
3 servings Fruit	½ cup prune juice; ½ banana for cereal
3 servings Fat	3 teaspoons margarine
1 Free Food	1 cup coffee

Lunch

3 servings Starches	2 slices whole-wheat bread
	¾ ounce unsalted pretzels
3 servings Meat	3 ounces sliced turkey
3 servings Fat	3 teaspoons mayonnaise
2 servings Vegetables	1 cup lettuce and tomato
	4 ounces V-8 vegetable juice
2 servings Fruit	1 medium apple
1 serving Dairy	1 cup skim milk

Dinner

5 servings Meat	5 ounces broiled red snapper
2 servings Vegetables	1 cup steamed broccoli
3 servings Starches	1 cup rice; 1 dinner roll
3 servings Fat	3 teaspoons margarine
1 serving Dairy	1 cup skim milk
3 servings Fruit	6 apricot halves (in their own juice)

Snack

4 servings Starches	4 cups air-popped popcorn
1 serving Fruit	½ cup orange juice

Daily Calorie Goal Chart: 2,000 Calories

Food Group	Number of Servings
Starches	10
Meat/Meat Alternatives	6
Dairy	3
Vegetables	4
Fruit	6
Fats	8
Free Foods	as desired

Daily Menu Example

These foods were chosen from our Food Groups lists appearing earlier in this chapter.

Breakfast

1 serving Dairy	1 cup skim milk for cereal and coffee
3 servings Starches	⅔ cup bran flakes
	2 slices whole-wheat toast
2 servings Fruit	½ banana for cereal
	½ cup orange juice
2 servings Fat	2 teaspoons margarine
1 Free Food	1 cup coffee

Lunch

2 servings Starches	2 slices whole-wheat bread
2 servings Meat	2 ounces sliced turkey
3 servings Fat	3 teaspoons mayonnaise
2 servings Vegetables	1 cup lettuce and tomato
	4 ounces V-8 vegetable juice
2 servings Fruit	1 medium apple

Dinner

4 servings Meat	4 ounces broiled red snapper
2 servings Vegetables	1 cup steamed broccoli
3 servings Starches	1 cup rice; 1 dinner roll
3 servings Fat	3 teaspoons margarine
1 serving Dairy	1 cup skim milk
2 servings Fruit	4 apricot halves (in their own juice)

Snack

2 servings Starches	2 whole graham crackers
1 serving Dairy	1 cup skim milk

Daily Calorie Goal Chart: 1,500 Calories

Food Group	Number of Servings
Starches	7
Meat/Meat Alternatives	6
Dairy	2
Vegetables	3
Fruit	4
Fats	6
Free Foods	as desired

Daily Menu Example

These foods were chosen from our Food Groups lists appearing earlier in this chapter.

Breakfast

1 serving Dairy	1 cup skim milk for cereal and coffee
3 servings Starches	⅔ cup bran flakes
	2 slices whole-wheat toast
1 serving Fruit	½ cup orange juice
2 servings Fat	2 teaspoons margarine
1 Free Food	1 cup coffee

Lunch

2 servings Starches	2 slices whole-wheat bread
2 servings Meat	2 ounces sliced turkey
2 servings Fat	2 teaspoons mayonnaise
1 serving Vegetables	1 cup lettuce and tomato
1 serving Fruit	12 grapes
1 Free Food	iced tea with lemon

Dinner

4 servings Meat	4 ounces broiled red snapper
2 servings Vegetables	1 cup steamed broccoli
3 servings Starches	½ cup rice; 1 dinner roll
3 servings Fat	3 teaspoons margarine
1 serving Dairy	1 cup skim milk
2 servings Fruit	4 apricot halves (in their own juice)

Snack

1 serving Starches	1 cup air-popped popcorn

Daily Calorie Goal Chart: 1,200 Calories

Food Group	Number of Servings
Starches	6
Meat/Meat Alternatives	5
Dairy	2
Vegetables	3
Fruit	2
Fats	4
Free Foods	as desired

Daily Menu Example

These foods were chosen from our Food Groups lists appearing earlier in this chapter.

Breakfast

1 serving Dairy	1 cup skim milk for cereal and coffee
2 servings Starches	⅔ cup bran flakes
	1 slice whole-wheat toast
1 serving Fruit	½ cup orange juice
1 serving Fat	1 teaspoon margarine
1 Free Food	1 cup coffee

Lunch

2 servings Starches	2 slices whole-wheat bread
2 servings Meat	2 ounces sliced turkey
2 servings Fat	2 teaspoons mayonnaise
2 servings Vegetable	1 cup lettuce and tomato
	4 ounces V-8 vegetable juice

Dinner

3 servings Meat	3 ounces broiled red snapper
1 serving Vegetables	½ cup steamed broccoli
1 serving Starches	½ cup rice
1 serving Fat	1 teaspoon margarine
1 serving Dairy	1 cup skim milk
1 serving Fruit	12 grapes

Snack

1 serving Starches	1 cup air-popped popcorn

This may come as a surprise to many of you, but it's now a well-accepted fact within the nutritional community. Researchers have been able to demonstrate the differing impact of fat and carbohydrate in the diet most dramatically in so-called induced-obesity studies. These are experiments that involve overfeeding healthy, normal-weight volunteers to make them gain body fat. They show that to gain X amount of body-fat weight, it takes five times as many calories from a mixed carbohydrate-fat diet than it does from a high-fat diet. In other words, *it's much harder to increase body fat on a high-carbohydrate diet than a high-fat diet.*

Why should this be the case?

The fat in food is more calorically dense. Your body extracts more than twice the number of calories from fat food versus carbohydrate and protein food. In other words, *our bodies react to fat as if it has more than double the number of calories, gram for gram, as carbohydrate or protein. Thus, food fat turns into body fat far faster than carbohydrate or protein food does.*

Your body has no trouble converting food fat into body fat, but it converts excess carbohydrate calories into body fat far less efficiently. Thus, if you want to shed body fat, you don't need to concentrate that much on counting carbohydrate and protein calories since they have little impact on your problem. But you should count fat-food calories assiduously, since they're the real culprits.

The disparity between the caloric impact of fats and carbohydrates/protein are shown in the box below, although we want to point out that some researchers have demonstrated an even greater disparity.[3] However, until their research gains wider currency, we prefer to use these, which are the traditional food composition values accepted by the majority of the nutritional community.

Metabolic Calorie Ratios

There are . . .

9 kcal in a gram of fat.

4 kcal in a gram of carbohydrate or protein.

Since fat is the real enemy, you need to know where it's lurking in order to avoid it. Figure 10-2 enables you to see at a glance the amount of fat in a representative sample of food. You may find some surprises on this list.

SPECIAL PHYSIOLOGICAL CHALLENGES FACING THE ELDERLY EATER

Much is made of elderly people's apathy about food. Some researchers view the problem purely as a psychosocial issue. Depression, a common problem with old people, is one of the most effective appetite suppressants ever devised. Moreover, many elderly people feel socially isolated and lonely. A person

Fat in a Representative Sample of Foods

	Grams of fat	Total calories	% Fat
MEAT, FISH, POULTRY, & EGGS			
Beef (3 oz trimmed of removable fat)			
corned beef	16	213	68
eye of round, roasted (select)	5	151	30
London broil, braised (choice)	12	208	52
porterhouse steak, broiled (choice)	9	185	44
rib, broiled (prime)	16	238	60
rib eye (Delmonico) steak, broiled (choice)	10	191	47
T-bone steak, broiled (choice)	9	182	45
top loin steak, broiled (select)	6	162	33
wedge-bone sirloin steak, broiled (choice)	8	180	40
Luncheon meats (1 slice)			
Louis Rich 96% Fat Free Turkey Pastrami	0	25	0
Louis Rich Oven Roasted Turkey Breast	0	30	0
Oscar Mayer Bologna	4	50	72
Oscar Mayer Hard Salami	3	35	77
Oscar Mayer 95% Fat Free Smoked Cooked Ham	1	25	36
Weaver Chicken Frank with Cheese	12	140	77
Seafood (3 oz cooked unless otherwise indicated)			
anchovies, canned in oil, drained, 5	2	42	43
Atlantic cod	1	89	10
haddock	1	95	9
lobster	1	83	11
salmon, pink, canned with bone and liquid	5	118	38
smoked salmon (lox)	4	99	36
swordfish	4	132	27
tuna, canned in oil and drained	7	158	40
tuna, canned in water and drained	0	111	0
shrimp	1	84	11
shrimp, breaded and fried	10	206	44
Poultry (3 oz roasted unless otherwise indicated)			
chicken breast, meat with skin	7	165	38
chicken breast, meat only	3	142	19
chicken drumstick, meat with skin, batter dipped and fried, 1 average	11	193	51
chicken drumstick, meat only, 1 average	2	76	24
chicken wing, meat only, 1 average	2	43	42
turkey, light meat with skin	7	168	38
turkey, light meat only	3	133	20
turkey, dark meat with skin	10	188	48
turkey, dark meat only	6	160	34
Eggs			
1 large	5	75	60
Fleischmann's Egg Beaters, ¼ cup	0	25	0
Morningstar Scramblers, ¼ cup	3	60	45

	Grams of fat	Total calories	% Fat
MILK & DAIRY PRODUCTS			
Milk *(1 cup)*			
whole	8	150	48
2% fat	5	120	38
1% fat	2	100	18
skim	0	90	0
buttermilk	2	99	18
Cream *(1 tbsp)*			
half and half	2	20	90
heavy whipping cream	6	52	100
sour cream	3	26	100
Cheese			
American, 1 oz	9	106	76
cheddar, 1 oz	9	114	71
cottage cheese, creamed, 1 cup	9	217	38
cottage cheese, 1% fat, 1 cup	2	164	11
cream cheese, 1 oz	10	99	91
mozzarella, 1 oz	6	80	68
mozzarella, part-skim, 1 oz	5	72	63
Parmesan, grated, 1 tbsp	2	23	78
ricotta, ½ cup	16	216	67
ricotta, part-skim, ½ cup	10	171	53
Swiss, 1 oz	8	107	67
Weight Watchers American Pasteurized Process Cheese Product, 1 slice	2	45	40
Yogurt			
Colombo, plain, 8 oz	7	150	42
Colombo, plain nonfat "lite," 8 oz	0	110	0
Dannon Fresh Flavors (coffee, lemon, vanilla), 8 oz	3	200	14
Dannon Original (all flavors), 8 oz	3	240	11
Yoplait Original (all flavors), 6 oz	3	190	14
FOOD FROM GRAINS			
Breads			
bagel, 1	1	163	6
English muffin, 1	1	135	7
whole-wheat bread, 1 slice	1	61	15
Cereals			
General Mills' Cheerios, 1¼ cups	2	110	16
Kellogg's Corn Flakes, 1 cup	0	100	0
Kellogg's Raisin Bran, ¾ cup	1	120	8
Nabisco Cream of Wheat, quick, 1 cup cooked	0	100	0
Nabisco Shredded Wheat, 1 biscuit	0	80	0
Old-Fashioned Quaker Oatmeal, ⅔ cup cooked	2	100	18
Quaker 100% Natural, ¼ cup	5	130	35
Quaker Puffed Rice, 1 cup	0	50	0

	Grams of fat	Total calories	% Fat
Crackers			
Sunshine Cheez-It Snack Crackers, 12	4	70	51
Nabisco Honey Maid Honey Grahams, 1 sheet	1	60	15
Nabisco Ritz, 4	4	70	51
Oyster crackers, 10	1	33	27
Other			
pasta, 1 cup cooked	1	159	6
white rice, 1 cup cooked	0	223	0
pancakes, 4″ plain	2	62	30
waffles, 7″ plain	8	206	35
French toast, 1 slice	7	153	41
Dunkin' Donuts Oat Bran Muffin, plain	11	350	28
Sara Lee Golden Corn Muffin	13	250	47
FRUITS & VEGETABLES			
apple, 1 medium	1	81	11
banana, 1 medium	1	105	9
fruit cocktail, canned in heavy syrup, ½ cup	0	93	0
orange, 1 medium	0	65	0
raisins, ⅓ cup	0	150	0
avocado, ½ medium	15	153	88
broccoli, ½ cup cooked	0	23	0
carrot, raw, 1 medium	0	31	0
corn, canned, ½ cup	1	66	14
green beans, ½ cup cooked	0	22	0
peas, ½ cup cooked	0	67	0
BEANS, NUTS, & SEEDS			
kidney beans, ½ cup boiled	0	113	0
lentils, ½ cup boiled	0	116	0
cashews, dry roasted, ¼ cup	16	197	73
coconut meat, dried, sweetened, flaked, ¼ cup	6	88	61
peanuts, dried, ¼ cup	18	207	78
peanut butter, 1 tbsp	8	95	76
pistachios, dry roasted, ¼ cup	17	194	79
sesame seeds, dried, ¼ cup	21	221	86
tahini (sesame butter), 1 tbsp	7	86	73
walnuts, black, dried, chopped, ¼ cup	18	190	85
SPREADS & OILS			
butter, 1 tsp	4	36	100
whipped butter, 1 tsp	3	27	100
margarine, stick & tub, 1 tsp	4	34	100
diet margarine, tub, 1 tsp	2	17	100

	Grams of fat	Total calories	% Fat
vegetable oil (corn, cottonseed, olive, peanut, canola, safflower, sesame, soybean, sunflower), 1 tbsp	14	120 (average)	100
vegetable oil spray, 2.5-second spray	1	6	100
SALAD DRESSINGS			
blue cheese, 1 tbsp	8	77	94
French, 1 tbsp	6	67	81
Italian, 1 tbsp	7	69	91
Russian, 1 tbsp	8	76	95
Thousand Island, 1 tbsp	6	59	92
SOUPS			
Campbell's Chicken Noodle, 1 cup	2	70	26
Campbell's Cream of Mushroom, 1 cup	7	100	63
Lipton Noodle, 1 cup	2	70	26
Progresso Green Split Pea, 1 cup	3	152	18
Progresso Beef Minestrone, 1 cup	3	135	20
Ramen Pride Oriental Noodles & Pork Flavor, 10 oz	8	198	36
SWEETS			
Cadbury's Milk Chocolate with Fruit & Nuts, 1 oz	8	150	48
Hershey Chocolate Kisses, 5 pieces	8	125	58
Snickers bar, 2.16 oz	14	290	44
Three Musketeers bar, 2.13 oz	9	260	31
angel food cake, 1/12 cake	0	126	0
brownie with nuts (3 x 1 x 7/8")	6	97	56
cheesecake, 1/8 cake	13	278	42
Hostess Ding Dong, 1	9	170	48
Hostess Twinkie, 1	5	160	28
pound cake, 1/2" slice	6	150	36
Almost Home Chocolate Chip Cookie, 2	5	130	35
Fig Newton, 1	1	50	18
Nabisco Gingersnaps, 4	3	120	23
apple pie, 1/8	12	282	38
banana cream pie, 1/8	12	233	46
pumpkin pie, 1/8	13	241	49
chocolate pudding, 1 cup	12	385	29
Dunkin' Donuts Plain Cake Ring	22	319	62
Nature Valley Granola Bar	5	120	38
Sara Lee Cheese Danish	8	130	55

	Grams of fat	Total calories	% Fat
FROZEN DESSERTS			
Ice cream			
Breyers, vanilla fudge twirl, ½ cup	8	160	45
Häagen Dazs, chocolate chocolate chip, ½ cup	18	290	56
Sealtest, vanilla, chocolate, or strawberry, ½ cup	6	140	39
Other			
Dole Fruit 'N Yogurt Bar	0	70	0
Dole fruit sorbet, strawberry, ½ cup	0	100	0
Eskimo Pie, 3 oz bar	12	180	60
Jell-O Chocolate Pudding Pop	2	80	23
Light n' Lively Ice Milk, heavenly hash, ½ cup	3	120	23
orange sherbet, ½ cup	3	92	29
Popsicle Ice Pop	0	50	0
Tofutti, wildberry supreme, ½ cup	12	210	51
Yoplait frozen yogurt, 3 fl oz	2	90	20
Toppings			
chocolate syrup, 2 tbsp	1	92	10
fudge topping, 2 tbsp	5	124	36
whipped cream, pressurized, 2 tbsp	1	16	56
DRY SNACK FOODS			
Lay's Bar-B-Q Flavored Potato Chips, 1 oz	9	150	54
Orville Redenbacher's Natural Microwave Popping Corn, 4 cups popped	7	110	57
popcorn, air-popped, 1 cup popped	0	23	0
Pringle's Light Potato Chips, 1 oz	8	150	48
pretzels, 1 oz	1	111	8
Ruffles Potato Chips, 1 oz	10	150	60

Source: Special Report, *Tufts Diet and Nutrition Letter,* 7, No. 8 (October 1989): 4–5.

who has always associated eating with sociability and now eats alone may simply lose interest. Some older people reject proffered food for symbolic reasons. It's an act of defiance aimed at caregivers, not to mention a way to vent the rage they may feel about their enfeebled condition.

While we don't mean to downgrade such realities, mental health isn't our bailiwick. Our focus is on the physiological reasons why some aging people between 65 and 75 years old begin to see food as more of a bother than a joy. There are real biological forces at work here, and they're just as detrimental as the emotional issues.

There are three prime challenges facing the older eater. They are: (1) a declining calorie requirement; (2) a flagging appetite; and

(3) a need to emphasize nutrients in the daily diet to prevent malnutrition. We'll discuss each and then tell you how to remedy the situation based on the latest research in our labs.

Challenge #1: A Declining Calorie Requirement

A declining metabolic rate is one of the Biomarkers you learned about back in Chapter 2. In our opinion, it falls in direct relation to older people's reduced muscle (lean-body) mass. With a reduced amount of muscle, the body's demand for oxygen during rest declines, as does its caloric need.

The daily energy needs of the average male and female fall over the course of a lifetime. People's caloric energy requirements peak when they're teenagers, when activity is frenzied, burning the candle at both ends is a way of life, and growth is rapid. Most teenagers eat enormous amounts of food. On average, boys eat up to 3,000 calories a day and girls, 2,200. Still, they tend to remain svelte and trim. Why? Because they have a high ratio of muscle to fat on their bodies, accompanied by a high metabolism and high activity.

But from about age 20 onward, things very gradually begin to change. As adults move toward middle age, their physical activity pace and metabolism slow, their muscle mass shrinks, and body fat grows. Their caloric requirements descend accordingly. The descent in daily energy needs becomes even more precipitous after age 50. Men over 50 years old need only about 2,300 calories a day to maintain a stable weight, while women need only 1,900.

For more insight into why active young people can stay slim while more sedentary older people fight the battle of the midriff, consider what happens to the food calories we humans eat:

- Our metabolism and lean-body mass determine what happens to about 60 to 70 percent of the calories we consume every day. Of course, when we've got a greater amount of fat on our body and a lower metabolic rate, we need less food calories to maintain the status quo, not to mention the fact that it's harder for us to lose weight.
- The amount of exercise or activity people undertake accounts for another 30 to 40 percent of food calorie intake. If a person is spending most of his or her waking hours sitting around, this means that person must eat correspondingly less to make sure the pounds don't start piling on.
- Finally, a very small amount of the calories we eat is burned

off immediately during the heat-generating (thermogenic) process of digesting and assimilating food nutrients.

This should help explain why a 30-year-old adult who is extremely active with a low body fat to lean-body mass ratio has a totally different caloric requirement from that of a 55-year-old person with a desk job, an addiction to TV, no inclination to exercise, and a weight problem. If you maintain a sedentary lifestyle as you age, your musculature is decreasing while body fat is accumulating, triggering a fall-off in your metabolic rate. This presarcopenic body composition and lower metabolism means your body needs fewer calories and expends what calories it gets more slowly.

Challenge #2: A Flagging Appetite

The appetite is one of the body's most miraculous adaptive mechanisms. An elaborate series of mechanisms controls eating behavior in normal adults. The body knows when to send signals of hunger or satiety to the brain to get us to reach for—or push away—more food. And it knows how to keep our internal metabolic engine chugging along steadily, even during those lengthy periods every day between meals when we're fasting.

It takes no more than 24 or 48 hours for the mechanisms that control our eating habits to recognize caloric overload—or deficiency. If we're eating too much and not exercising enough, our body notes the imbalance, then sends signals to encourage us to make compensatory changes in the volume of food we eat. Studies have shown that calorie intake and output in normal adults is usually in equilibrium over a period of 4 to 14 days.

As we move beyond middle age, however, there is some evidence that appetite, our body's ability to regulate calorie intake and expenditure, becomes less efficient. But that's by no means the whole story. There is also evidence that it's not only a decline in the appetite mechanism that's the problem, but certain related factors that older people cannot control. Foremost among them are deteriorating senses of taste and smell and the side effects of medications some older people must take for chronic conditions.

Taste and smell are largely responsible for the enjoyment we feel when we eat a delicious meal. The circumvallate papillae are structures in the tongue that house our "taste buds." Young adults have 250 taste buds in each papilla. But by the time many people reach their mid-70s, they have less than 100 buds per papilla, a striking reduction. In addition, the thresholds for all four primary tastes—salty, sweet, bitter, and sour—increase in older age groups. This means foods have to have stronger tastes for the elderly to get the same sensation.

What may have an even more deleterious effect on eating habits is the desensitization of the nose. Between youth and old age, the ability to smell all essences and flavors decreases almost ninefold. Thus, the concentration of a substance must be nine times greater in order for an older person to achieve the same smell sensation as a young person. Dr. Richard L. Doty, director of the Smell and Taste Center at the University of Pennsylvania Hospital, had done studies indicating that a quarter of people between 65 and 79 years old and half of those 80 and above have lost most of their sense of smell.[4]

Should you doubt the importance of your olfactory ability in making food more palatable, try this experiment. Blindfold yourself and place nose plugs or a clothespin over your nose. Have someone spoon-feed you a meal without telling you what you're eating. You'll be surprised how few foods you'll be able to identify with any certainty. This will give you a firsthand idea of the deprivation many very old people suffer and why eating often strikes them as a chore.

Then there's the problem of medications. Drugs may alleviate the symptoms of many degenerative diseases, but they can also interfere with food intake by affecting the appetite, either directly or by altering a person's sense of smell and taste. If you or an older person you know is taking a drug regularly, read the fine print about possible side effects.

If you're a middle-aged person who is valiantly fighting the battle of the bulge, it may seem like a blessed relief to contemplate a time when you will no longer have your present-day craving to eat, eat, eat. Don't be misled. A poor appetite in your later years is not a good thing for this reason:

The less people eat, the harder it becomes for them to eat a well-rounded diet, replete with the essential nutrients they need to stay healthy. Less food being eaten means there's less margin for error in the selection of food.

Challenge #3: Increasing Beneficial Nutrients in the Daily Diet

To be sure, malnutrition is a problem for some of the elderly. In our experience at Tufts monitoring elderly people's eating habits, we find it takes an especially vigilant person, ever conscious of the importance of good nutrition, to choose the right foods in the right proportions day in and day out. When you're consuming smaller amounts of food every day, there's small margin for error.

On the other hand, we also concede that malnutrition in the elderly has many causes beyond a slack appetite, boredom, and depression. Other very real problems in the elderly are the change in their body's ability to absorb certain nutrients and detrimental drug interactions.

Older people have a marked decrease in vital stomach acid secretions, a phenomenon known as "atrophic gastritis." This "silent," asymptomatic change affects one-quarter of those over 65 and up to 40 percent of all 80-year-olds. Stomach acid secretion helps to release nutrients from food and to foster the assimilation process that takes place in the intestine. With a lessened ability to make stomach acid, some nutrients pass through older people's bodies without being fully utilized. The nutrients most likely to be affected are vitamin B_{12}, folic acid, iron, and calcium.

Medication enters the picture here, too. Many therapeutic drugs have nutrient-blocking side effects. Drugs can block the body's absorption ability and nutrient metabolism, but it's rarely for more than one or two nutrients, fortunately. Or a drug might change the process of nutrient excretion. The risk of such adverse nutritional side effects from medications increases as a person takes more drugs simultaneously or takes one drug over a long period of time.

MEETING THE PHYSIOLOGICAL CHALLENGES FACING THE ELDERLY EATER

The first two challenges—a declining calorie requirement and flagging appetite—have the same remedy: regular exercise.

Yes, we are once again beating the drum for exercise. But we have to, for exercise is still the world's best natural appetite stimulant no matter what your age. It's the antidote for a waning appetite—whether you're 80 or 104.

Whether you're middle-aged or elderly, regular exercise is the missing link. Here are the formulas for good health for these two age groups:

• *In your middle years,* your appetite is strong. To assuage it without depriving yourself, you need to eat selectively—but not necessarily that much less—while you're exercising regularly in order to expend the maximum amount of calories every day. Your goal is to retain—or regain—a body high in muscle and low in fat.

• *In old age,* when sarcopenia looms as a threat, you're still exercising to maintain a body high in muscle and low in fat. You're also exercising to maintain your appetite, knowing that exercise will foster hunger to meet the body's need for more energy to propel working muscles. Your dietary goal is to eat a variety of healthful foods because, for older people, simply eating more goes a long way toward eating right.

RESEARCHING THE MATURE PERSON'S DIET

Having said all this, you'd assume our next tack would be to outline a diet to combat aging. Given the present state of our scientific knowledge about nutrition and aging, this is not as easy as it seems.

What we know for sure is still sketchy. At least we know what we don't know and need to probe further. Here are a few of our discoveries at Tufts that may influence the publication of the next Recommended Dietary Allowances (RDAs).[5]

If we could do a long-term study tracking the amounts of nutrients circulating in the blood as well as some other measurements of nutritional status over the course of a person's lifetime, no doubt we would find age-related declines that parallel those in our 10 physiological Biomarkers. We suspect such age-related declines can be altered, even reversed, just as we've found ways to influence the course of our 10 physiological Biomarkers. Some of these interventions might involve a dietary change only. Others might require a combination exercise-diet solution.

• For example, we know that the calcium balance in the body is disrupted with age owing at least partially to a decline in the body's ability to absorb calcium. To treat the problem, we need to do several things:

One is simply to make sure older people get enough calcium in their diet in forms that their bodies can readily absorb. It's an indisputable fact that elderly women who have too little calcium in their diet lose mineral from their spine at a much greater rate than women who take more than the RDA for that nutrient. On the other hand, we find that calcium supplements, when taken with a meal, can impair iron absorption in the food just eaten. One dietary solution involves an increase in vitamin D intake, which aids in calcium absorption.

Of course, to maintain adequate bone mineral content in the elderly, there's also the exercise remedy. Studies have shown pretty conclusively that weight-bearing exercise (walking and jogging)—both alone and in conjunction with calcium supplements—helps maintain the vitality of the bone-density Biomarker.

• As with calcium, vitamin D status declines with age. We've seen this in studies in various older populations. Nursing home residents have especially low levels. The ability of the skin to synthesize vitamin D from sunlight decreases markedly with age, not to mention the fact that many of the elderly get less exposure to the sun.

Finding ways to maintain appropriate vitamin D levels in older people is a very promising area. Again, this may be a case where the elderly's dietary requirement for this nutrient should be increased. We're even wondering if vitamin D can help improve older people's muscle strength and function, one of our Biomarkers. We're doing further studies to find out.

• Early research findings seem to indicate there may be some correlation between cataract formation, a major problem in the elderly, and vitamin C intake. Cataracts, which cause blurred vision, are due to an accumulation of damaged proteins in the lens of the eye. Cataracts are rare in younger people because normal processes in their eyes clear out any protein excess.

We know that one reason for cataracts is accumulated damage from exposure to the ultraviolet rays, mostly from the sun. We're now learning that such antioxidant nutrients as vitamin E, carotene, and vitamin C seem to afford some protection from ultraviolet light. In laboratory animals, increasing the amount of vitamin C in the lens of the eye diminishes the tendency for cataract formation. We've found that older people who have low amounts of vitamin C in their blood are at higher risk of forming cataracts.

With all we're learning about cataracts, we even think there may come a time when we can elevate older people's failing vision to the status of a Biomarker. By our definition, a Biomarker is a declining biological function that you can influence for the better through a lifestyle change. In this case, the only change required might be a dietary one.

• We've found a provocative link between vitamin C and HDL-cholesterol levels. However, we need to examine the relationship further before we even arrive at any preliminary hypothesis.

• We didn't include the decline in older people's perceptual and neurological abilities in our list of Biomarkers because we don't have enough solid evidence yet on how we can intercede to halt the slide. However, this is one instance where a dietary/nutritional intervention may eventually turn out to be beneficial.

We know that the amounts of vitamins B_6 and B_{12} in the body decline with age. In the case of B_{12}, we found this is largely due to that decrease in stomach acid secretions we mentioned earlier. Studies to date show that B_{12} absorption decreases in at least one-third of the elderly who have lost the ability to make stomach acid. Stomach acid is necessary for full absorption of both dietary vitamin B_{12} and another vitamin, folic acid.

We now have methods that enable us to measure the presence of mild vitamin B deficiencies and study their symptoms. As a consequence, we've seen that deficiencies in vitamins B_6, B_{12}, and folic acid can trigger such neurological changes as a drop in alertness and memory ability as well as numbness and tingling in the legs.

Some of our newest research focuses on homocysteine, an amino acid circulating in the blood that is a sensitive indicator of mild vitamins B_6, B_{12}, and folic acid deficiency. Elevated homocysteine levels indicate a person is suffering from such a deficiency. In recent surveys, up to 15 percent of the elderly had elevated homocysteine levels, which went back to normal when the mild deficiencies were corrected. Here's the revealing part: With the deficiencies gone, some subjects showed improvement in mental or neurological function. This suggests that the B_{12} shot our grandmothers got from that old country doctor was more than just a placebo in many instances.

In light of these findings and with more research, we may come to the conclusion that the current vitamin B_6 and vitamin B_{12} Recommended Dietary Allowances for the elderly should be higher. No, we do not think that the maintenance of vitamin B in the elderly means that all memory loss, Alzheimer's disease, and senile dementia has a nutritional cause and can be corrected by that means. We do think it means we may one day be able to prevent a small percentage of such ailments through changes in dietary habits and nutritional status.

• Our studies show that it may be harder for the elderly to get enough riboflavin (vitamin B$_2$) and zinc in their diets. Riboflavin is a key nutritional helpmate because it helps release energy from carbohydrates, proteins, and fat. It also aids in the maintenance of the body's mucous membranes. Too little zinc can also pose problems for the elderly, lessening their appetite, diminishing their sense of taste, and lengthening the time it takes for wounds to heal.

• Immunity—our body's ability to ward off infectious diseases as well as fight tumor growth and the like—declines as we grow older. Indeed, some investigators think the fall-off in our immune function may be central to the whole process of aging.

In older people, T lymphocytes, which figure heavily in immunity, seem to have a lessened ability to respond to stimulation, to swing into action when needed, as it were. In addition, we know that malnourished people, both children and adults, have a marked decrease in their immune function, leading one to question whether there's a relationship between declining nutritional function and declining immune function as we grow older.

At the Human Nutrition Research Center on Aging, we've shown that supplementing elderly people's diet with vitamin E, an antioxidant nutrient, for one month improves their immune responsiveness. In our experiments we've also found that vitamin E appears to play multiple roles in the bodies of the elderly. Given these findings, we suspect the requirement for vitamin E and/or other dietary antioxidants, such as vitamin C, may have to be reconsidered. What we need now is more conclusive evidence that they can counteract some of the age-related decline in the immune system. Indeed, what we've learned so far makes us wonder if immunity may also achieve Biomarker status one day. Is it an example of a declining function we may be able to alter through diet?

• Our investigation to date indicates that an elderly person's need for vitamin A may actually decrease with age, meaning the RDA for an older population may need lowering. In one of our studies, we found that the elderly are much slower to clear vitamin A from the blood. They are more likely to store it longer in tissue and have a higher level of a form of vitamin A that is associated with toxicity. The complication is that carotene, found in many vegetables, especially carrots, is an important nutrient for the elderly. Once inside our body, our liver converts carotene into vitamin A. So we have to contend with the very tricky issue of how to separate vitamin A needs from carotene needs.

DIETARY ADVICE FOR THE ELDERLY EATER: MORE AND BETTER FOOD, *NOT* MORE SUPPLEMENTS

If you've passed the 60 mark and you're concerned about meeting special nutrient needs, we urge you to begin by focusing on changing your eating habits. Include in your daily menu more of the foods shown in the table below. Do not simply buy bottles of supplements containing these nutrients. *Supplements cannot convert a careless diet into a good one.* Our goal in this chapter is to get you to understand more about basic nutrition, thereby enabling you to make wiser food choices. It is not to see you line your pantry shelves with bottles of vitamin and mineral pills.

Nutrients to Emphasize in the Aging Person's Diet★

Vitamins	Food Sources
Vitamin B_6	liver, herring, salmon, nuts, brown rice
Vitamin B_{12}	meat, liver, kidney, egg yolk, fish
Folic Acid	liver and kidneys, dark-green leafy vegetables such as spinach, wheat germ, dried peas and beans
Vitamin C	black currants, sweet peppers, broccoli, citrus fruits
Vitamin D	milk, liver oils, tuna, salmon, herring, egg yolk, margarine
Vitamin E	vegetable oils, wheat germ oil, olives, peanut oil
Minerals	
Calcium	milk, cheese, dairy food, sardines
Zinc	meat, eggs, liver, seafood
Magnesium	unrefined cereals
Chromium	whole grains, leafy green vegetables

★ In the Food Group lists earlier in this chapter, the foods with asterisks are also good sources of the nutrients elderly people need more of.

The average middle-aged person, by careful selection and preparation of food, can get adequate amounts of the right nutrients, including those featured in the table. Supplementation usually isn't necessary; it should be a last resort. For some people over 70 years old, supplementation may make sense. When supplements are used, the nutrients in the table should be present in any multivitamin or multimineral preparation you choose. For individual recommendations tailored to unique needs, the advice and counsel of a physician or trained, registered dietitian is helpful.

RATS AREN'T PEOPLE

We don't want to leave this subject without commenting briefly on the long series of caloric-restriction studies done on rats over the last fifty years. Those of you given to reading books on longevity or the health column of your daily newspaper are undoubtedly familiar with them and aware of the findings.

Back in 1935, C. M. McCay and his colleagues were the first investigators to observe that rats fed consistently less calories than their counterparts over the course of their lifetime lived longer.[6] Everything about these moderately starved rats' lives was elongated when they were compared with their well-fed peers. It was as if their restricted diet had triggered some invisible switch that threw their lives into slow motion. The aging process in these ultralean rodents was much slower and more deliberate. It took them far longer than average to mature physiologically, reach senescence, and finally die.

Since McCay's time, there's been a long list of follow-up studies, almost all of which replicated his results. The consistency of the findings has led some medical commentators to draw an analogy that strikes us as somewhat strained. They equate rats allowed to eat at will with overfed, sedentary businesspeople, while they compare the caloric-restricted rats to those people who eat sensibly with restraint their whole lives.

We concede that this phenomenon is still something of a puzzlement. Gerontologists have devised various theories to explain it, but none strike us as the whole story. About the only thing we think these studies say definitively is that nutrition can play a crucial role in the aging process.

Human beings, exposed to all sorts of environmental and psychological stresses during their lives, are certainly not caged rats. Nor are we at all sure that the aging mechanisms operative for rodents are identical with those in larger animals, especially primates. In fact, evidence from population studies in humans suggests the opposite of the rodent results. In Chapter 2 you learned that people with an especially low or high body-mass index appear to have a shorter life expectancy. For humans, we certainly don't recommend overeating, but we're not advocates of semistarvation, either. We urge you to beware claims that have grown out of these rodent experiments.

We realize a well-balanced daily menu sounds pretty dull compared to a brand new diet or health-food concoction, and that people truly hooked on such approaches are difficult to budge. We also find that people who are swayed by nutritional fads are often those who had very poor eating habits earlier in their lives. Now that they suddenly recognize the link between nutrition and health, they feel that a radical approach is the only thing that can adequately right their past wrongs. To their way of thinking, complete retribution in the form of a punitive eating program is the only pathway to nutritional salvation.

Untrue. *Eating a normal, sensible diet at any age can only help. It will help reverse any damage done by previous poor eating habits just as much as—and probably more than—any extreme diet ever could.*

MAKING YOUR HEALTH SPAN
MATCH YOUR LIFE SPAN

The late B. F. Skinner, the famous behavioral psychologist, made one of the more reasoned arguments we've heard for getting middle-aged people to plan for their physical old age. From the perspective of someone who had achieved a hardy seniority himself, he wrote in his book *Enjoy Old Age:*

> If you were planning to spend the rest of your life in another country, you would learn as much about it as possible. You would read books about its climate, people, history, and architecture. You would talk with people who had lived there. You might even learn a bit of its language

> Old age is rather like another country. You will enjoy
> it more if you have prepared yourself before you go.[1]

Skinner concludes that "a good time to think about old age is
when you are young, because you can then do much to improve
the chances that you will enjoy it when it comes."

Extending Skinner's metaphor, we'd like you to think of this
Afterword as a kind of travel guide to the country of old age.
That country may be a long way off, but we all know we're going
to visit it one day. Whether we're fortunate enough to travel there
with a spouse or loved one at our side or we make the trip alone,
it's one journey none of us can escape. As a serious tourist, it's
smart to know what to expect so you can plan for it accordingly.

Here we'd like to dispel some common misconceptions you
may harbor about old age, and assure you that what lies ahead is
far brighter than you may think.

MEDICINE CAN KEEP YOU WELL—NOT JUST MAKE YOU WELL

We think people make a grievous mistake when they view
doctors simply as wonder-working technicians whose job it is to
alleviate bodily ills *after the fact*. The doctor-as-technician model is
counterproductive in our society because it gives too many people
the false impression that they no longer have to practice those
healthy-living principles we all learned back in hygiene classes in
elementary and high schools. We were all taught the connection
between the quality of the food we eat, our physical activity level,
and long-term health. Thus, it's quite astonishing that, as adults,
so many of us overlook the whole concept of preventive medicine. Instead we equate our bodies with pieces of equipment that
break down periodically and need to go to the shop for tinkering.
If we get sick, we're confident some miracle-working physician
will always be around to "fix" the problem, much like our
friendly local garage mechanic fixes our car. We assume that as
long as our money—or insurance—holds out, we can just keep
getting our deteriorating body repaired far into the future.

This approach is foolhardy, especially when it comes to the
diseases associated with advancing age. The fact is that the degenerative diseases of old age—cardiac conditions, cancer, diabetes,
hypertension, osteoarthritis, osteoporosis, hypertension, hearing

and visual impairment, and so on—are *chronic diseases*. They rarely go away completely. Drugs and other medical therapies can't cure them as they can often cure *acute, infectious diseases*. Chronic disease symptoms may disappear for a while, it's true. But never mistake that for a cure. The underlying malady is still in residence and will surely recur again and again, usually at the most unexpected and inconvenient moments. To be sure, these diseases are no respecters of anyone else's timetable.

Given what we know and still don't know right now about the causes of degenerative ailments, we feel that preventive strategies, such as our Biomarkers Program, hold as much promise as many medical therapies for improving health in the later years.

The notion we articulate in our Biomarkers Program is basically the same one emerging from the so-called wellness movement spearheaded in the 1980s by private-sector employers in league with farsighted health-care institutions. Our idea, and theirs, is to educate the public to the notion that . . .

You are largely responsible for your own health throughout your whole life, even well into old age.

LIFE EXPECTANCY IS INCREASING—THE HUMAN LIFE SPAN IS NOT

Because of all the marvelous medical discoveries of our time —kidney and heart transplants, antibiotics, and the full panoply of life-saving and life-extending techniques discovered in this century—many people are under the impression that the human life span is increasing. This isn't true. Life expectancy is increasing.

• "Life expectancy" pinpoints the age when the average person, living in a specific country, can expect to die given that society's environmental conditions; infant mortality, accident, and disease rates; health-care system, and so on. Life expectancy rates vary greatly from one country to the next and between the industrialized and developing regions of the world.

Life expectancy has changed enormously down through the ages. In ancient Greece life expectancy at birth was 22 years, which is not to say plenty of people didn't live longer.[2] For an American child born in 1776, it was 35 and had moved up only four years a century later.[3] By 1900 Americans were doing a little better; life expectancy at birth was 47.

From then until now, the progress in Americans' life expectancy has been nothing short of remarkable. Today the U.S. life expectancy figure hovers around 75, a little less for men and a little more for women. A whole cluster of factors are responsible —everything from fewer deaths during childbirth for mothers and lower infant mortality to better housing, indoor plumbing, immunizations, and antibiotics. How much of the credit can be claimed by medical advances? Less than you'd think.

Today Japan has the highest life expectancy in the world: 75.2 years for men and 80.9 years for women.[4] This may not be surprising since Japan is one of the most advanced nations in the world. It may surprise you to learn, though, that the life expectancy rates in most developing countries now surpass the rates of 1900 in the United States and the other industrialized nations.

• "Life span" is an altogether different matter. It's a biological term referring to the age at which the average member of a species would die if there were no diseases, predators, or accidents to cause a premature demise. This, of course, is a purely theoretical scenario since these things are always present in one form or another whether the species in question is man or beast.

To date, 114 years is the oldest verifiable age that a person has lived, although there are anecdotal accounts galore—that cannot be documented—of people who have lived well into their fifteenth or sixteenth decade.

Yes, we know all about the books and articles extolling the long lives of peoples in remote mountainous regions of the USSR, Pakistan, and Ecuador. Stories abound about such people living to ripe old ages of 150 or more. These tales are particularly cherished by anthropologists and journalists. Many have journeyed to these far-off climes and lived briefly among these latter-day Methuselahs searching for clues about why they live so long when those of us in so-called civilized places don't.

For all the accounts that support these claims of long life, we can cite just as many treatises—usually written by more credentialed authors who used better methods of investigation—that dispute them.

James Fries and Lawrence Crapo, two professors at the Stanford University School of Medicine, reviewed all the accounts and wrote up their findings in an amusing essay called "The Myth of Methuselah."[5] Among other things, they stress the dearth of verifiable records, the close correlations between regional illiteracy and the percentage of claimed centenarians in a society, the overwhelming number of male claimants when women are known to live longer in most other places in the world, and the odd fact that Christian populations have fewer centenarians than Moslems. Could it be that the Moslem ten-month year has caused a misunderstanding?

They conclude that "age exaggeration seems as prevalent in the very old, *particularly the poorly educated,* as is age reduction for the middle-aged."

If humans only grew rings the way trees do, maybe we could settle these longevity claims once and for all. Unfortunately, without them, or some similar kind of definitive evidence, these controversies will continue to rage.

What is instructive about these Shangri-la tales is that *there are important lifestyle patterns common to the regions that harbor the allegedly long-lived.*

By our standards, the societies in question are hardship cases. They're rural, mountainous, and in high altitudes where the air is thin; moreover, visitors are rare and incomes are very low. Granted, few of us enjoying the conveniences of life in modern industrialized countries would want to trade places, even if it did mean a longer life. But hardship isn't the end of the story. There are positive things about these longevous people's lives that we "civilized" folk could—and *should*—emulate to a far greater degree than we do now.

What are they?

Fries and Crapo—and other respected investigators such as Alexander Leaf[6]—point out that these long-lived people have *"diets low in calories and animal fats.* There is a *strikingly high level of physical activity* and fitness, including active farming and tilling the ground in old age. *Obesity is extremely uncommon.* There is *moderation in alcohol and tobacco consumption.* Very importantly, there is *no retirement* and elderly people remain active in social and economic life; a sense of usefulness and purpose pervades the lives of old people."

The fact that the aged in these traditional cultures retain, even increase, their social status certainly holds a lesson for us in countries where the elderly are put out to pasture and largely ignored —and where middle-aged citizens tend to be overburdened, overweight, time-bound, exercise-starved, frustrated, and unduly stressed.

It's interesting that American researchers, probing why some of our senior citizens manage to outlive the life expectancy figures despite the hassles of modern life, have turned up evidence along similar lines.

Erdman Palmore, a driving force behind Duke University's landmark study on aging, studied 268 volunteers (aged 60 to 94) over a period of years to determine what factors were the best predictors of their longevity. He concluded there were four: The best was each person's actuarial life expectancy at initial testing, which is another way of saying our life expectancy figures are pretty accurate. The second best predictor was a person's physical health. Third was their work satisfaction. And fourth was their scores on standard intelligence tests.[7]

Utilizing about the same number of volunteers but new statistical and analytical methods, Palmore arrived at roughly the same conclusion several years later: "The most important overall factors in longevity among the aged are health maintenance (especially avoiding cigarettes); and maintaining a useful and satisfying role and positive view of life."[8]

A more recent study of some 7,000 residents of Alameda County, California, in the San Francisco Bay Area turned up similar results. Investigators Lester Breslow and Nedra Bellow identified a handful of simple guidelines for a longer life. They are no smoking, weight within 20 percent of that recommended for one's age, sex, and height; moderate alcohol intake (which they found was actually better than either total abstinence or abuse); moderate exercise about three times a week; regular meals, including breakfast; and sleeping seven or eight hours a night.

Perhaps their most startling finding was that the _70-year-olds who followed all these health rules were as healthy as people, aged 35 to 44, who practiced only three!_ Breslow and Bellow concluded that this antiaging regimen was responsible for the male study participants living, on average, an extra eleven years and the females living an extra seven.[9]

THE IDEAL: A LONG HEALTH SPAN AND SHORT DISABILITY SPAN

Think of your adult life divided into two periods. The first period is your heath span, which stretches over most of your life and is characterized by vigorous activity, general physical well-being, and complete self-reliance. The second period, at the end of a life, is characterized by illness, invalidism, and total dependence, hence the term *disability span,* or what we call the "Disability Zone."

Statisticians who study disease and mortality figures say that *people are becoming older and older when chronic infirmities first strike.* (See box below.) Since, as we said earlier, the human life span is fixed at around 114, the older people are when they become incapacitated, the shorter their disability span can last.

Health vs. Disability Span Expectancy for Americans, Ages 65–69

	Men	Women
Remaining Total Life Expectancy	13.1 years	19.5 years
Remaining Active Life Expectancy ("Health Span")	9.3 years	10.6 years
Remaining Dependent Life Expectancy ("Disability Span")	3.8 years	8.9 years

Source: Hamish N. Munro, "Aging and Nutrition: A Multifaceted Problem," *Nutrition and Aging,* eds. M. L. Hutchinson and H. N. Munro (Orlando, Fl.: Academic Press, 1986).

A short *disability span* at the end of a vigorous and generous *health span* is the situation our Biomarkers Program seeks to foster. As the table above indicates, it's a scenario that's already becoming more the norm than the exception in the United States as we approach the turn of the next century.

Here's where wellness regimens like our Biomarkers Program can exert a powerful impact. By following such a program for the rest of their lives, middle-aged people can greatly improve their chances of approaching the ideal: a health span that almost matches their life span.

Resources

DETERMINING DISEASE RISK VIA YOUR WAIST-TO-HIP RATIO

To assess where fat is distributed on a person's torso, scientists use a measurement called the "waist-to-hip ratio." This measurement targets fat storage just above and below the waist. It involves measuring the waist, in a relaxed state, and dividing this measurement by your hip measurement. For example, if your waist measures 30 and your hips 48, your waist-to-hip ratio is .625.

Here's how to do it yourself:

- Using a fabric tape measure, the kind used for sewing, measure your waist at the level of your navel or "belly button." *Do not* suck in your stomach or take a deep breath when doing this. Stay relaxed.
- Next, measure the circumference around your buttocks or hips. Do it at the place where the protrusion is the greatest.

- It doesn't matter whether you take these measurements in inches or centimeters. Either way, divide your waist/navel measurement by your hip/buttocks measurement. The figure you get is your waist-to-hip ratio.

The interpretive charts (below) explain what your waist-to-hip ratio means in terms of your risk for developing coronary artery disease or diabetes. Using the chart for your sex, find your age along the bottom and your waist/hip ratio along the left grid. Trace a line straight up from your age and across from your waist/hip ratio. Where the two lines intersect, you'll find your risk category.

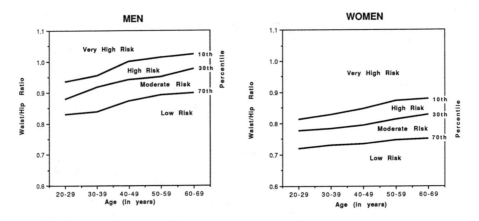

Compare the two charts for a moment. Two things should be evident:

The first is that men generally have a higher waist-to-hip ratio than women. Men's fat is usually deposited *around the waistline* in the form of potbellies, or "barrel chests." In contrast, women's fat tissue tends to collect *below the waistline,* contributing to the term *hourglass figure.*

However, this does not mean that women are at less risk. Rather, as the graphs show, when a woman deviates from the norm for her sex and has even slightly more fat around and just above her waist, she moves into a higher risk category much sooner than a man does.

APPENDIX B

AEROBIC SELF-ASSESSMENT: WALK-FOR-TIME TEST SCORING CHARTS

Chapter 3 contains a walking test to measure aerobic capacity. Once you've taken the test and have figured out your heartbeats per minute, your next step is to consult the appropriate chart for your age and sex in this appendix.

Here's how you use the chart:

On the bottom, put a dot indicating how long (in minutes) it took you to complete the walking course. Then trace a line straight up to the place on the diagonal grid indicating your heartbeats per minute and put another dot there. To ascertain your maximum aerobic capacity, simply draw a straight line to the left edge of the chart.

The figure you arrive at tells you nothing until you compare it with the norms on the interpretive fitness table back in

Chapter 3. This table will tell you how your aerobic capacity stacks up to that of your chronological peer group.

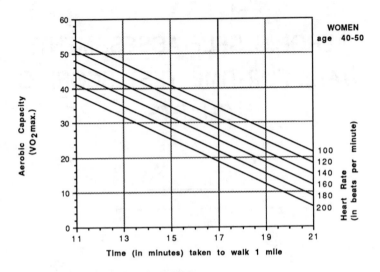

Chart B1 **WALK-FOR-TIME TEST:**
SCORING CHART FOR WOMEN, AGE 40–49

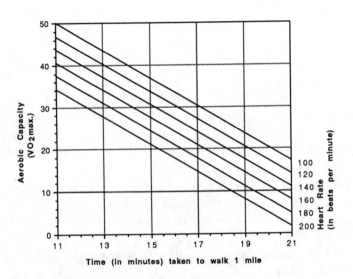

Chart B2 **WALK-FOR-TIME TEST:**
SCORING CHART FOR WOMEN, AGE 50–59

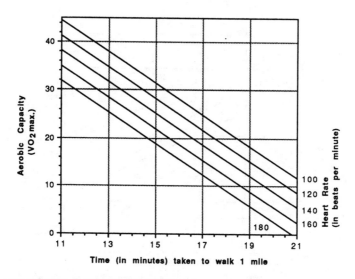

Chart B3 **WALK-FOR-TIME TEST:**
SCORING CHART FOR WOMEN, AGE 60–69

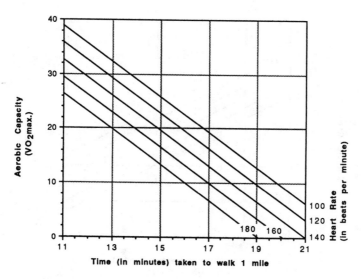

Chart B4 **WALK-FOR-TIME TEST:**
SCORING CHART FOR WOMEN, AGE 70–79

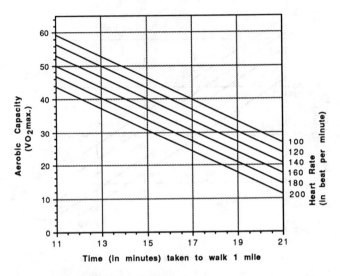

Chart B5 **WALK-FOR-TIME TEST:**
SCORING CHART FOR MEN, AGE 40–49

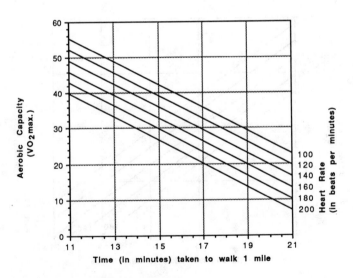

Chart B6 **WALK-FOR-TIME TEST:**
SCORING CHART FOR MEN, AGE 50–59

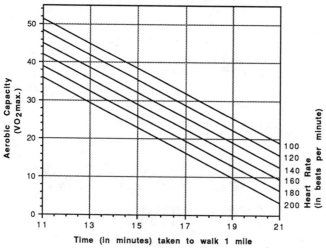

Chart B7 WALK-FOR-TIME TEST:
SCORING CHART FOR MEN, AGE 60–69

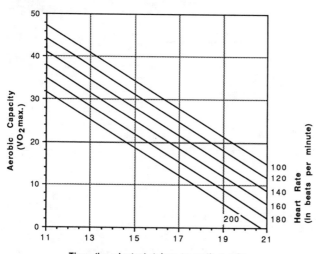

Chart B8 WALK-FOR-TIME TEST:
SCORING CHART FOR MEN, AGE 70–79

The charts were derived from the following formula:
VO_2max. $= 132.853 - (0.0769 \times \text{weight}) - (0.3877 \times \text{age}) + (6.31 \times \text{sex}) - (3.2649 \times \text{time}) - (0.1565 \times \text{heart rate})$

Here's the key:
Weight is in pounds.
Age is in years.
Male = 1
Female = 0
Heart rate is for a complete minute.

Time is in minutes to the neare
fraction (in other words, 15
minutes and 15 seconds =
15.25).

Source: J. M. Rippe, A. Ward, J. P. Porcari, and P. S. Freedson, "Walking for Health a Fitness," *Journal of the American Medical Association* 259 (1988): 2720–24.

AEROBIC SELF-ASSESSMENT ALTERNATIVE: THE STEP TEST *

In this section we describe an alternative way to test aerobic capacity indoors. This test—somewhat more complicated than the outdoor Walk-for-Time Test contained in Chapter 3—measures your heart rate in response to a stepping exercise, which you'll be doing three consecutive times at three different speeds.

Once again, you'll be taking your pulse (see figure 3-1) in order to assess your performance. You need a sturdy stool that's about a foot high, a chair placed about three feet behind the stool, and a metronome. The metronome available in music supply stores enables you to specify a precise rhythm or exercise intensity

---

* This Step Test was developed by Herbert deVries and Dianne Hales and described in their book, *Fitness After 50* (New York: Charles Scribner's Sons, 1984).

and then to reproduce that same rhythm at some later date when you retest yourself. (In lieu of a stool, you could use the bottom step of a flight of stairs, provided the step is high enough. In lieu of a metronome, you need a helpmate—a spouse or friend—who can mark time like a metronome with the aid of a watch with a sweep second hand.)

This test is complicated and will require a dress rehearsal of sorts to make sure you thoroughly comprehend all that's involved. Even if you own a metronome, we still suggest that you do the test with the aid of a partner who can double- check everything you've done to make sure you get it right. Study the instructions carefully before you begin.

In preparation for the test, take your pulse at rest and write it here:

My resting pulse rate is _____ beats per minute.

If you're fortunate enough to have a metronome, find the following three speeds on it:

The first speed corresponds to 4 ticks every 5 seconds.

The second involves 4 ticks every 3½ seconds.

The third, 4 ticks every 2½ seconds.

If you don't own a metronome, your partner must act as a human metronome and count aloud, "One . . . two . . . three . . . four," while consulting a watch to make sure the speed is right.

The test is divided into three one-minute phases with recovery periods in between when you sit down and count your pulse steadily for two minutes.

Phase 1

• With the metronome—or your partner clocking 4 ticks per 5 seconds, you step up on the stool and down from it 12 times. In one minute's time you'll actually be taking 48 steps, one step for each metronome tick. Those steps, and precise instructions, are shown here:

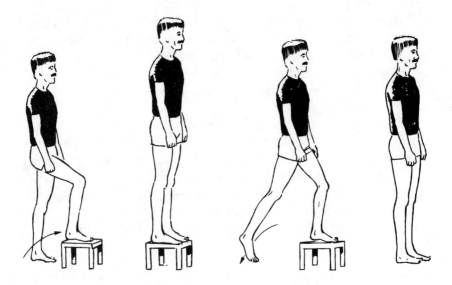

- At the end of the minute, your partner signals you to stop and sit down. You immediately begin counting your pulse for two minutes. Your partner signals when the two minutes are over. *Write all the beats you count here:* _____
- Continue to sit there until your pulse returns to its resting rate of _____ beats per minute, then immediately embark on Phase 2.

Phase 2

- With the metronome (or partner) now clocking the faster speed of 4 ticks every 3½ seconds, you'll step up and down off the stool a total of 18 times—or 72 steps corresponding to each metronome tick.
- At the end of the minute, your partner once again signals you to halt, be seated, and count your pulse for another two minutes. *Write all the beats you count here:* _____.
- Rest on the chair only until your pulse returns to its resting rate of _____ beats per minute.

Phase 3

- During the final phase, the metronome—or your partner's voice—is counting at a very brisk clip of 4 ticks every 2½ seconds—and you're stepping on and off that stool just as smartly. You step up and down a total of 24 times, in the process taking a total of 96 steps corresponding to each metronome tick.

- When the minute is up, you sit down and count your pulse for a final two minutes. *Write all the beats you count here:* _____.

Step Test Scoring

This is a graduated test that forces your heart to work harder during each successive phase. That's why the number of heartbeats you count during each phase gets progressively larger.

On the following charts in the column for your age group, circle the range of numbers corresponding to your results:

Step Test: Phase 1 Total Heartbeats in Two Minutes

Score	Age 50–59	Age 60–69	Age 70–79
Excellent	102–120	108–125	111–128
Good	121–140	126–145	129–148
Average	141–158	146–162	149–165
Below Average	159–198	163–202	166–205

Step Test: Phase 2 Total Heartbeats in Two Minutes

Score	Age 50–59	Age 60–69	Age 70–79
Excellent	110–126	115–130	118–133
Good	127–146	131–151	134–154
Average	147–166	152–170	155–173
Below Average	167–206	171–210	174–214

Step Test: Phase 3 Total Heartbeats in Two Minutes

Score	Age 50–59	Age 60–69	Age 70–79
Excellent	114–130	118–134	122–137
Good	131–150	135–154	138–157
Average	151–170	155–174	158–167
Below Average	171–210	175–214	178–217

To score yourself, transfer the results above to the scorecard on the next page.

Step Test Scorecard ★

Phase 1 score:	Phase 2 score:	Phase 3 score:
Excellent	Excellent	Excellent
Good	Good	Good
Average	Average	Average
Below Average	Below Average	Below Average

• If the majority of your circles are around "Excellent," you're an exceptional human specimen and, we suspect, a serious athlete.

• If most of your circles are distributed in the "Excellent" and "Good" categories, you fall into the classification of moderate aerobic fitness.

• If your circles are clustered in the "Average" and "Below Average" designations, you fall into the category of low aerobic fitness.

★ Transfer these scores to the composite fitness scorecard at the end of Chapter 3.

Step Test Precautions

You're right if this test strikes you as physically challenging. For most people, it is. Because the test will push some of you toward the limits of your endurance, some precautionary words are in order:

If you feel too exhausted after Phase 1 or 2, don't continue the test. Or, if you find you can't keep up with the metronome speed, do not proceed to the next phase; instead, score yourself in the Below Average category.

Maybe, with the utmost difficulty, you do manage to get through all three phases. However, when you score yourself you notice that you drop down from an "Excellent" or "Good" rating in Phase 1 to a "Below Average" rating in Phase 2 or 3. In other words, you drop down two or more categories between Phases 1 and 3. This is alarming. It indicates an inappropriate cardiovascular response to exertion and you should talk to your doctor about it. Describe exactly what you did and how your pulse reacted. Do not proceed with the Biomarkers Program until you've had a thorough checkup and secured medical clearance.

GLOSSARY:
THE BIO-VOCABULARY OF AGING

This glossary will acquaint you with the terms associated with the latest thinking and insight into aging.

Aerobic capacity This is your body's ability to use oxygen to produce energy within a given time. As people age, the capacity for oxygen use in a given time usually declines, a situation that can be reversed through regular exercise. (See also "VO$_2$max.")

Blood–sugar tolerance (or glucose tolerance) This refers to our body's ability to control the amount of blood sugar (or glucose) in our blood after eating or drinking a defined amount of sugar. Blood-sugar levels tend to rise with age, although impaired glucose tolerance, like high blood pressure, has no symptoms. Our research indicates that this age-related blood-sugar impairment is most likely caused by the increasingly sedentary lifestyle of the average aging person and a diet too high in fat. Poor glucose tolerance can lead to non–insulin-dependent diabetes mellitus (NIDDM).

BMR (basal metabolic rate) The rate of our body's chemical processes when it's at rest. The BMR tends to decline with age, largely because the amount of muscle tissue in people's bodies tends to decline with advancing age, a situation that can be reversed through regular exercise.

Body composition This is a very general term that describes how much fat, muscle, bone, and so forth, there is in our bodies. Using the most traditional techniques for measuring body composition, we can assess the amount of fat and nonfat (also called "lean-body mass") tissue. Aging is associated with increasing amounts of fat and decreases in lean-body mass, specifically muscle.

Body-fat mass   The body's less metabolically active storage tissue, technically known as "adipose tissue." With advancing age, even if our body weight doesn't change much, most of us tend to gain fat and lose heavier muscle tissue.

Cholesterol/HDL ratio   Cholesterol is a fatty substance that circulates in the bloodstream in association with protein in entities known as "lipoproteins." HDL (high-density lipoprotein) is a beneficial component of cholesterol that acts as a kind of scouring agent, cleansing the arteries of plaque, which, in excess, can cause fatal cardiovascular obstructions. While our HDL-cholesterol tends to remain constant, our total cholesterol level tends to increase as we age, an indication that the dubious LDL (low-density lipoprotein) or VLDL (very low-density lipoprotein) components are increasing. Both correct diet and regular exercise can reverse such imbalances.

Dehydration   Dehydration simply means having a lower amount of body water than normal. Aging tends to be associated with chronic dehydration due to age-related decreases in kidney function and thirst. The result is a reduction in the body's vital thermoregulatory ability.

Lean-body mass   Muscle, bone, and other vital organ tissue of the body—in short, everything that is *not* fat. This is the crucial component of the body that our Biomarkers Program seeks to preserve and build back up.

Life expectancy   This is a statistical estimate of how long a person can expect to live based on such variables as his or her year of birth, infant mortality, societal disease rates, sex, and environment. Unlike the human "life span" (see below), which is fixed, life expectancy changes. For example, Americans in 1900 had a life expectancy at birth of 47. A baby born in the year 2000 will have a projected life expectancy in the United States of over 77 years. A 92-year-old nursing home patient has a remaining life expectancy of about one year.

Life span   Every living species has a distinct life span. For dogs, it's 13 to 17 years depending on size; for elephants, around 70 years; for Texas rattlesnakes in captivity, 16 years. For human beings, it's around 100, give or take a decade or two.

Despite all the effort and research dollars that have been poured into life-extension research in recent years, science has done nothing to alter the human life span. Science, in tandem with various environmental and lifestyle factors, however, has been remarkably successful at helping us to die closer to the end of our natural, species-specific life span and to have a longer health span and shorter disability span.

Muscle   The largest component of lean-body mass. Ideally, it should be the largest component of our entire body composition. It is responsible for vitality and health to a much greater extent than most people realize. It can be increased and strengthened through exercises like those we feature in this book.

One repetition maximum (1RM)   The most weight a person can lift with one try. Our Biomarkers Program asks people to do strength-building exercises with weights or resistance rubber bands that are 80 percent of their 1RM.

Osteoporosis   An age-related condition, resulting from the gradual loss of bone mineral content, that increases risk of bone fracture. Osteoporosis is characterized by weaker, less dense, more brittle bones. Loss in bone mineral may be affected by diet and exercise.

Sarcopenia   A term, first used in this book, that refers to an overall weakening of the body caused by a gradual, decades-long change in body composition, with loss of muscle mass.

Strength   The maximum amount of force that your muscles can generate. The ability to increase muscle strength through muscle-conditioning exercises does not appear to be diminished with age.

VO$_2$max.   A scientific symbol that stands for the maximum amount of oxygen a person can take in during heavy exertion. VO$_2$max. is measured in laboratories by asking subjects to breathe into a special apparatus while they exercise to the point of exhaustion on a treadmill or a stationary bike. Older people have to exercise regularly over a longer period of time to achieve VO$_2$max. levels equivalent to those of younger adults.

CHAPTER NOTES

Introduction

[1]George Burns, *How to Live to Be One Hundred or More* (New York: G. P. Putnam's Sons, 1983), p. 131.

[2]Bengt Saltin, G. Blomqvist, J. H. Mitchell, R. L. Johnson, K. Wildenthal, and C. B. Chapman, "Response to Submaximal and Maximal Exercise After Bedrest and Training," *Circulation* 38 (1968), Supplement 7.

[3]Walter M. Bortz II, "Disuse and Aging," *Journal of the American Medical Association* 248, No. 10 (September 10, 1982): 1203–8.

Chapter 1

[1]George Burns, *Dear George: Advice and Answers from America's Leading Expert on Everything from A to B* (New York: G. P. Putnam's Sons, 1986).

[2]Steven N. Blair, H. W. Kohl, R. S. Paffenbarger, D. G. Clark, K. H. Cooper, and L. W. Gibbons, "Physical Fitness and All-Cause Mortality: A Prospective Study of Healthy Men and Women," *Journal of the American Medical Association* 262 (November 3, 1989): 2395–2401.

Chapter 2

[1]Alex Comfort, *The Biology of Senescence,* 3rd ed. (New York: Elsevier, 1979).

[2]Maynard J. Smith, "Review Lectures on Senescence, I: The Causes of Aging," *Proceedings of the Royal Society of London,* Series B, 157 (1962): 115–127.

[3]D. Rudman et al., "Effects of Human Growth Hormone in Men Over Sixty Years Old," *New England Journal of Medicine* 323 (1990): 1–6.

[4]L. Larsson, G. Grimby, and J. Karlsson, "Muscle Strength and Speed of Movement in Relation to Age and Muscle Morphology," *Journal of Applied Physiology* 46 (1979): 451–56.

[5]T. Moritani and H. A. deVries, "Potential for Gross Muscle Hypertrophy in Older Men," *Journal of Gerontology* 35 (1980): 672–82. In this training study, the researchers used an imprecise measurement of muscle gain following weight training to come to the erroneous conclusion that hypertrophy failed to occur and that any strength gains resulted from "neural factors" such as the brain "learning" to recruit more muscle cells.

[6]W. R. Frontera, C. N. Meredith, K. P. O'Reilly, H. G. Knuttgen, and W. J. Evans, "Strength Conditioning in Older Men: Skeletal Muscle Hyper-

trophy and Improved Function," *Journal of Applied Physiology* 64 (1988): 1038–44.

[7]M. A. Fiatarone, E. C. Marks, N. D. Ryan, C. N. Meredith, L. A. Lipsitz, and W. J. Evans, "High-Intensity Strength Training in Nonagenarians: Effect on Skeletal Muscle," *Journal of the American Medical Association* 263 (1990): 3029–34.

[8]Body-mass index (BMI) is calculated, arithmetically, by dividing your weight in kilograms (1 kg = 2.2 lbs.) by the square of your height in centimeters (1 centimeter = .394 inches). In equation form, it looks like this:

$$\frac{\text{Weight (in kilograms)}}{\text{Height} \times \text{Height (both in centimeters)}} = \text{Your body mass index (BMI)}$$

An example: A man who weighs 180 pounds weighs 82 kilograms (that's 180 pounds divided by 2.2 = 81.818 kilograms). If he's 6 feet (72 inches) tall, his height is 183 centimeters (72 inches divided by .394 = 182.7 centimeters). When you plug these numbers into the equation, we find the man has a BMI of .0024, or 24.

[9]R. Andres, D. Elahi, J. D. Tobin, D. C. Muller, and L. Brant, "Impact of Age on Weight Goals," *Annals of Internal Medicine* 103 (1985): 1030–33.

[10]R. P. Donahue, R. D. Abbott, E. Bloom, et al., "Central Obesity and Coronary Heart Disease in Men," *Lancet* 8537 (1987): 821–24.

[11]R. E. Ostlund, Jr., M. Staten, W. Kohrt, J. Schultz, M. Mallery, "The Ratio of Waist-to-Hip Circumference, Plasma Insulin Level, and Glucose Intolerance as Independent Predictors of the HDL (sub 2) Cholesterol Level in Older Adults," *New England Journal of Medicine* 322 (1990): 229–34.

[12]Roy J. Shephard, "Gross Changes of Form and Function," *Physical Activity and Aging,* 2nd ed. (Rockville, Md.: Aspen Publishers, 1987), pp. 93–94.

[13]D. M. Seals, J. M. Hagberg, B. F. Hurley, A. A. Ehsani, and J. O. Holloszy, "Endurance Training in Older Men and Women, 1: Cardiovascular Responses to Exercise," *Journal of Applied Physiology* 57 (1984): 1024–29. Also C. N. Meredith, W. R. Frontera, E. C. Fisher, V. A. Hughes, J. C. Herland, J. Edwards, and W. J. Evans, "Peripheral Effects of Endurance Training in Young and Old Subjects," *Journal of Applied Physiology* 66 (1989): 2844–2849.

[14]E. P. Reaven and G. M. Reaven, "Structure and Function Changes in the Endocrine Pancreas of Aging Rats with Reference to the Modulating Effects of Exercise and Caloric Restriction, *Journal of Clinical Investigation* 68 (1981): 75–84.

[15]J. W. Anderson and N. J. Gustafsson, "Type II Diabetes: Current Nutrition Management Concepts," *Geriatrics* 41 (1986): 28–38.

[16]N. F. Gordon and L. W. Gibbons, *The Cooper Clinic Cardiac Rehabilitation Program* (New York: Simon & Schuster, 1990), p. 195.

[17]D. Streja and D. Mymin, "Moderate Exercise and High-Density Lipoprotein Cholesterol," *Journal of the American Medical Association* 243 (1979): 2190–2192. See also M. M. Dehn and C. B. Mullins, "Physiologic Effects and Importance of Exercise in Patients with Coronary Artery Disease," *Cardiovascular Medicine* 4 (April 1977): 31–47.

[18]A. S. Truswell, B. M. Kehnelly, J.D.L. Hansen, and R. B. Lee, "Blood Pres-

sure of !Kung Bushmen in Northern Botswana," *American Heart Journal* 84 (1972): 5–11.

[19]S. N. Blair, N. Goodyear, and L. W. Gibbons, "Physical Fitness and the Incidence of Hypertension in Healthy Normotensive Men and Women," *Journal of the American Medical Association* 252 (1984), 487–95.

[20]F. L. Kaufman, R. L. Hughson, and J. P. Schaman, "Effect of Exercise on Recovery Blood Pressure in Normotensive and Hypertensive Subjects," *Medicine and Science in Sports and Exercise* 19 (1987): 17–20.

[21] B. Dawson-Hughes, P. F. Jacques, and C. N. Shipp, "Dietary Calcium Intake and Bone Loss from the Spine in Healthy Postmenopausal Women," *American Journal of Clinical Nutrition* 46 (1987), 68–87.

[22]B. Krolner and B. Toft, "Vertebral Bone Loss: An Unheeded Side Effect of Therapeutic Bed Rest," *Clinical Science* 64 (1983): 537–540.

[23]M. E. Nelson, C. N. Meredith, B. Dawson-Hughes, and W. J. Evans, "Hormone and Bone Mineral Status in Endurance-Trained and Sedentary Postmenopausal Women," *Journal of Clinical Endocrinology and Metabolism* 66 (1988): 927–33.

[24]M. E. Nelson, E. C. Fisher, and W. J. Evans, "A One-Year Walking Program and Increased Dietary Calcium in Postmenopausal Women: Effects on Bone," *Medicine and Science in Sports and Exercise* 22 (Supplement, 1990): 377.

[25]E. L. Smith, W. Reddan, and P. E. Smith, "Physical Activity and Calcium Modalities for Bone Mineral Increase in Aged Women," *Medicine and Science in Sports and Exercise* 13 (1981): 60–64.

[26]P. A. Phillips et al. "Reduced Thirst After Water Deprivation in Healthy Elderly Men," *New England Journal of Medicine* 311 (1984): 753–59.

Chapter 5

[1]William J. Evans, "Exercise-Induced Skeletal Muscle Damage," *The Physician and Sportsmedicine* 15 (January 1987): 89–100.

Chapter 7

[1]Steven N. Blair, H. W. Kohl, R. S. Paffenbarger, D. G. Clark, K. H. Cooper, and L. W. Gibbons, "Physical Fitness and All-Cause Mortality: A Prospective Study of Healthy Men and Women," *Journal of the American Medical Association* 262 (November 3, 1989): 2395–2401.

[2]The American College of Sports Medicine recently added strength training to their official position paper on how much and what kind of exercise adults need in order to stay fit and healthy. See the April 1990 issue of *Medicine and Science in Sports and Exercise,* the official ACSM journal.

Chapter 8

[1]Frederick A. Whitehouse, "Motivation for Fitness," *Guide to Fitness After Fifty,* eds. Dr. Raymond Harris and Lawrence J. Frankel (New York: Plenum, 1977), pp. 171–89.

[2]R. E. Vartabedian and K. Matthews, *Nutripoints* (New York: Harper & Row, 1990), p. 433.

[3]Kenneth H. Cooper and Mildred Cooper, *The New Aerobics for Women* (New York: Bantam Books, 1988), p. 75.

Chapter 9

[1]An exception is the longitudinal study of M. L. Pollack et al., "Effect of Age and Training on Aerobic Capacity and Body Composition of Master Athletes," *Journal of Applied Physiology* 62 (1987): 725–31.

[2]Jack H. Wilmore and David L. Costill, *Training for Sport and Activity,* 3rd ed. (Dubuque: William C. Brown, 1988), pp. 147–48.

[3]D. L. Costill, *Inside Running: Basics of Sports Physiology* (Indianapolis: Benchmark Press, 1986).

[4]Wilmore and Costill, *Training for Sport and Activity,* pp. 147–48.

[5]Ibid., p. 195.

[6]Wm. J. Evans and Virginia A. Hughes, "Dietary Carbohydrates and Endurance Exercise," *American Journal of Clinical Nutrition* 41 (May 1985): 1146–54.

[7]C. N. Meredith, M. J. Zackin, W. R. Frontera, and W. J. Evans, "Dietary Protein Requirements and Body Protein Metabolism in Endurance-Trained Men," *Journal of Applied Physiology* 66 (1989): 2850–56.

Chapter 10

[1]M. Peterson and K. Peterson, *Eat to Compete* (Chicago: Year Book Medical Publications, 1988), pp. 279–80.

[2]Ibid., pp. 279–80.

[3]Dr. Elliot Danforth of the University of Vermont School of Medicine has done extensive studies on the effects of carbohydrate and fat calories in the body. Based on his research, the disparity between fat and carbohydrate calories is far greater than the traditional 9 to 4 differential. Historically, nutritionists have assumed that once our calorie needs are met, the excess carbohydrate we've consumed is converted to fat tissue. Dr. Danforth finds this is not entirely true for the following reasons: Our bodies store fat as adipose tissue —that is, in cells specialized to hold fat. To a much more limited extent, our bodies also store carbohydrate for use as quick energy. When we eat carbohydrates—bread, pasta, cereal, etc.—the sugar molecules are extracted from those foods and used to replace any depletion in our small carbohydrate stores and to fuel tissue metabolism. Any excess carbohydrate calories are seldom converted and stored as adipose tissue simply because the task is too difficult. Our bodies would have to muster an awful lot of caloric energy to effect such a conversion, technically called "lipogenesis." The lipogenesis process, in and of itself, expends lots of carbohydrate calories, leaving little left over for body fat storage. Moreover, according to Danforth, our bodies may not have enough of the enzymes primed for lipogenesis to do a thorough job of converting all the excess carbohydrate we've eaten into body fat. In short, our capacity for lipogenesis is limited. Consequently, fewer excess carbohydrate calories than expected, based on pure calorie counts, actually end up as fat. (For similar reasons, the same can be said about the consumption of excess protein calories, which are not a significant contributor to body fat either.) Danforth concludes that it takes 23 percent of ingested carbohydrate calories to pay the metabolic cost of

lipogenesis, while food fat can be converted into body fat utilizing only 3 percent of the ingested fat calories. That's a huge sevenfold difference.

[4]Jane E. Brody, "For Millions, Little or No Sense of Smell Means Diminished Life Quality and Special Problems," _New York Times,_ February 1, 1990, B10.

[5]In the United States, Recommended Dietary Allowances (RDAs) are the accepted standards of nutrition, published periodically by the National Academy of Sciences. Since 1974 there has been a separate category of RDAs for people 51 years of age or older, but the recommendations are largely identical with those for young adults. Moreover, lumping active people in their fifties and sixties with the frail elderly in their seventies and eighties simply does not make a lot of sense. Ideally, we'd like to see RDA categories of 25–59, 60–74, 75–84, and 85 and over, because they'd more accurately reflect the heterogeneity of the adult population. But this didn't happen in the most recent 1989 RDA pronouncement, _Recommended Dietary Allowances,_ 10th ed., compiled by the Food and Nutrition Board (part of the National Research Council) and published by National Academy Press (Institute of Medicine, 2101 Constitution Avenue, N.W., Washington, D.C. 20418). The next edition isn't due out until 1994.

[6]C. M. McCay, M. F. Crowell, and L. A. Maynard, "The Effect of Retarded Growth upon the Length of the Life-Span and upon Ultimate Body Size," _Journal of Nutrition_ 10, (1935): 63–79.

Afterword

[1]B. F. Skinner and M. E. Vaughan, _Enjoy Old Age: A Program of Self-Management_ (New York: W. W. Norton, 1983), p. 20.

[2]R. B. Greenblatt, "Aging Through the Ages," _Geriatrics_ 32(6) (June 1977): 101–02.

[3]U.S. Census Bureau, _Historical Statistics of the United States: Colonial Times to 1970_ (Washington, D.C.: Government Printing Office, 1971).

[4]Ken Dychtwald, _Age Wave_ (Los Angeles: Jeremy P. Tarcher, 1989), p. 10.

[5]James F. Fries and Lawrence M. Crapo, _Vitality and Aging: Implications of the Rectangular Curve_ (San Francisco: W. H. Freeman, 1981), 11–18.

[6]Alexander Leaf, "The Aging Process: Lessons from Observations in Man," _Nutrition Review_ (February 1988): 8.

[7]Erdman Palmore, "Physical, Mental, and Social Factors in Predicting Longevity," _Normal Aging: Reports from the Duke Longitudinal Study, 1955–1969_ (Durham, N.C.: Duke University Press, 1970), pp. 406–16.

[8]Erdman Palmore, "Predicting Longevity: A New Method," _Normal Aging, II: Reports from the Duke Longitudinal Study, 1970–1973_ (Durham, N.C.: Duke University Press, 1974), p. 289.

[9]L. Breslow and J. E. Enstrom, "Persistence of Health Habits and Their Relationships to Mortality," _Preventive Medicine_ 9 (1980): 469–83. See also N. Belloc, "Relationship of Health Practices and Mortality," _Preventive Medicine_ 2 (1973): 67–81.

INDEX

———

ABOUT THE AUTHORS

IRWIN H. ROSENBERG, M.D., has been the Director of the USDA Human Nutrition Research Center on Aging at Tufts University since 1986. He is a physician and physiologist whose special interest is clinical nutrition and metabolism. He is former Chairman of the Food and Nutrition Board of the National Academy of Sciences and former President of the American Society of Clinical Nutrition, and has held medical professorships at Harvard and the University of Chicago. He lives in Boston, Massachusetts.

WILLIAM J. EVANS, PH.D., is the Chief of the Human Physiology Laboratory at the HNRCA. His research on the physiology of aging and sports performance has been widely published in the medical literature and publicized in the lay media. He is a Fellow of the American College of Sports Medicine as well as the American College of Nutrition, and Exercise Adviser to the Boston Bruins and the New England Patriots. Dr. Evans lives in Acton, Massachusetts.

JACQUELINE THOMPSON is a best-selling writer in the areas of business, fitness, and health. She lives in Staten Island, New York.